T-Labs Series in Telecommunication Services

Series editors

Sebastian Möller, Berlin, Germany
Axel Küpper, Berlin, Germany
Alexander Raake, Berlin, Germany

More information about this series at http://www.springer.com/series/10013

Fatemeh Ganji

On the Learnability
of Physically Unclonable
Functions

 Springer

Fatemeh Ganji
Institut für Softwaretechnik
    und Theoretische Informatik
Technischen Universität Berlin
Berlin
Germany

ISSN 2192-2810   ISSN 2192-2829   (electronic)
T-Labs Series in Telecommunication Services
ISBN 978-3-030-09563-5   ISBN 978-3-319-76717-8   (eBook)
https://doi.org/10.1007/978-3-319-76717-8

Printed on acid-free paper

This Springer imprint is published by the registered company Springer International Publishing AG
part of Springer Nature
The registered company address is: Gewerbestrasse 11, 6330 Cham, Switzerland

Dedicated to my mother, my sister, and the memory of my father.

List of Publications Related to this Thesis

The primary results presented in this thesis have been first published in the following papers:

- **Ganji, F.**, Tajik, S., Fäßler, F., Seifert, J.P.: Having No Mathematical Model May Not Secure PUFs. Journal of Cryptographic Engineering, Springer-Verlag, 2017
- **Ganji, F.**, Tajik, S., Fäßler, F., Seifert, J.P.: Strong Machine Learning Attack Against PUFs with No Mathematical Model. In Proceedings of 18th International Conference on Cryptographic Hardware and Embedded Systems— CHES 2016, Santa Barbara, USA
- **Ganji, F.**, Tajik, S., Seifert, J.P.: Let Me Prove It to You: RO PUFs Are Provably Learnable. In Proceedings of 18th Annual International Conference on Information Security and Cryptology (ICISC), 2015, Busan, South Korea
- **Ganji, F.**, Tajik, S., Seifert, J.-P.: Why Attackers Win: On the Learnability of XOR Arbiter PUFs. In Proceedings of 8th International Conference on Trust and Trustworthy Computing—TRUST 2015, Heraklion, Greece
- **Ganji, F.**, Tajik, S., Seifert, J.P.: PAC Learning of Arbiter PUFs. Journal of Cryptographic Engineering, Springer-Verlag, 2016

In addition to the aforementioned papers, Fatemeh Ganji has co-/authored the following publications:

- **Ganji, F.**, Krämer, J., Seifert, J.P., Tajik, S.: Lattice Basis Reduction Attack against Physically Unclonable Functions. In Proceedings of 22nd ACM Conference on Computer and Communications Security—CCS 2015, Denver, USA
- Tajik, S., Lohrke, H., **Ganji, F.**, Seifert, J.P., Boit, C.: Laser Fault Attack on Physically Unclonable Functions. In Proceedings of Workshop on Fault Diagnosis and Tolerance in Cryptography (FDTC), IEEE, 2015, St. Malo, France

Contents

1 Introduction 1
  1.1 Motivation 1
     1.1.1 Hardware Root of Trust 2
     1.1.2 Fragile Security of ICs 2
  1.2 Physically Unclonable Functions 3
  1.3 Thesis Statement 5
     1.3.1 Problem Statement 5
     1.3.2 Our Attack Model 6
     1.3.3 Thesis Contributions 7
  1.4 Outline of the Thesis 8

2 Definitions and Preliminaries 9
  2.1 Notations 9
  2.2 PUFs 9
  2.3 Boolean Functions 11
     2.3.1 Linearity of Boolean Functions 12
     2.3.2 Average Sensitivity of Boolean Functions 13
     2.3.3 Non-linearity of PUFs over $\mathbb{F}_2$ and the Existence
            of Influential Bits 14
  2.4 Linear Threshold Functions 15
  2.5 Regular Language and Principles of DFAs 17
  2.6 Probably Approximately Correct Model 18

3 PAC Learning of Arbiter PUFs 21
  3.1 Introduction 21
  3.2 Representing Arbiter PUFs by DFAs 23
     3.2.1 Discretization Process of Delay Values 25
     3.2.2 Building a DFA Representing an Arbiter PUF 26
  3.3 PAC Learning of Arbiter PUFs 29
  3.4 Comparison with Related Work 32

   3.5   Practical Considerations 33
       3.5.1   The Important Role of M 33
       3.5.2   Dealing with the Metastable Condition 34

4   PAC Learning of XOR Arbiter PUFs 35
   4.1   Introduction 36
   4.2   LTF Representation of XOR Arbiter PUFs.................. 37
   4.3   PAC Learning of XOR Arbiter PUFs 39
   4.4   PAC Learning of Noisy XOR Arbiter PUFs................. 43
   4.5   Discussion 44
       4.5.1   Theoretical Considerations 44
       4.5.2   Practical Considerations 46

5   PAC Learning of Ring Oscillator PUFs 49
   5.1   Introduction 50
   5.2   DL Representation of RO-PUFs 51
   5.3   PAC Learnability of the $2 - DL$ Representing the RO-PUF 53
   5.4   Results and Discussion 55

6   PAC Learning of Bistable Ring PUFs 59
   6.1   Introduction 60
   6.2   Architecture of the BR-PUF Family....................... 61
   6.3   A Constant Upper Bound on the Number of Influential Bits 62
       6.3.1   Heuristic Approaches............................ 62
       6.3.2   A Boolean-Analytical Approach 64
   6.4   PAC Learning of PUFs Without Prior Knowledge of Their
       Mathematical Model.................................. 66
   6.5   Results and Discussion 68

7   Follow-Up Work 73
   7.1   Lattice Basis Reduction Attack 73
   7.2   Laser Fault Injection Attack 74

8   Conclusion and Future Work 77

References 81

Acronyms

AdaBoost	Adaptive Boosting
BR	Bistable Ring
CRP	Challenge-Response Pair
DFA	Deterministic Finite Automata
DL	Decision List
DT	Decision Tree
EM	ElectroMagnetic
FPGA	Field-Programmable Gate Array
GHz	Gigahertz
IC	Integrated Circuit
IoT	Internet of Things
IP	Intellectual Property
LR	Logistic Regression
LTF	Linear Threshold Function
ML	Machine Learning
NIST	National Institute of Standard and Technology
nm	Nanometer
PAC	Probably Approximately Correct
PDF	Probability Density Function
PLD	Programmable Logic Devices
ps	Picosecond
PUF	Physically Unclonable Function
RO	Ring Oscillator
RoT	Root of Trust
SAC	Strict Avalanche Criterion
TBR	Twisted Bistable Ring
UART	Universal Asynchronous Receiver/Transmitter (UART)
XOR	Exclusive-OR

Symbols and Operators

Symbols

$\overline{\alpha}_{i,j}$	The realizations of the partial sums $\mathcal{A}_i$ at the outputs of the ith stage of an arbiter PUF, where $j = 1$ for upper and $j = 2$ for lower output. $\overline{\alpha} \in \mathbb{R}$
$\alpha_{i,j}$	An integer-valued delay associated with $\overline{\alpha}_{i,j}$
$\overline{\beta}_{i,j}$	With $1 \leq i \leq 4$ and $1 \leq j \leq 2$, the realizations of the variables $\mathcal{B}_i$ in an arbiter PUF
$\beta_{i,j}$	An integer-valued delay associated with $\overline{\beta}_{i,j}$
γ	The precision of the arbiter terminating the chain in an arbiter PUF; furthermore, it denotes the error of a weak PAC learning algorithm
δ	The confidence level of a PAC learning algorithm
$\delta(\cdot, \cdot)$	The transition function associated with a DFA
Δ	The delay difference at the outputs of the last stage in an arbiter PUF
ε	The accuracy level of a PAC learning algorithm
η	Noise rate
η_b	Upper bound on the noise rate η
λ	The empty string of length $|\lambda| = 0$
μ_i	The mean of the random variables $\mathcal{B}_i$
φ	With $1 \leq i \leq n$ and $0 \leq j \leq 1$, the delay difference at the output of the ith stage
Φ	The set of all $\mathbf{\Phi}$'s
$\mathbf{\Phi}$	The input vector of a Perceptron; when learning XOR Arbiter PUF, it is the encoded challenge vector
$\chi_S(c)$	A Fourier character that equals $(-1)^{\sum_{i \in S} c_i}$, when $c \in \mathbb{F}_2$
σ	The minimum distance of any example from $\mathcal{P}$

σ_i	The deviation of the random variables $\mathcal{B}_i$
Σ	The alphabet $\Sigma = \{0, 1\}$
Σ^*	The set of all strings over Σ
ω	The weight vector used to generate the output of the Perceptron
$\Omega(\cdot)$	Big Omega notation
$\mathcal{A}_i$	The random variable corresponding to the total delay between the enable point and the outputs of the ith stage of an arbiter PUF
$\mathcal{B}_i$	The random variable related to the delay within the ith stage of an arbiter PUF
c	A challenge applied to a PUF; a Boolean string
c_i	The ith bit of the challenge, or the ith variable of a Boolean string
$\mathbf{c}_i$	ith challenge vector corresponding to the ith CRP collected from a PUF
$c^{\oplus i}$	Obtained by flipping the ith bit of the challenge
$\mathcal{C}$	The set of the challenges that can be applied to a PUF
C_n	The set of all binary strings with n bits
d_i	Deviation of an example given to the Perceptron algorithm
$\deg(f)$	The degree of the Fourier expansion of a function f
$\deg_{\mathbb{F}_2}(f)$	The degree of the $\mathbb{F}_2$-polynomial representation of a Boolean function f
D	$D = \sqrt{\sum_{i=1}^{r} d_i^2}$; furthermore, in PAC model, it refers to an arbitrary distribution
EX	The Oracle
EX_η	Noisy Oracle
$\mathbb{F}_2$	The finite field of size 2. Here, we mainly refer to $\mathbb{F}_2 = \{0, 1\}$
f_i	A term of maximum size k in a k-DL
$\hat{f}(S)$	The Fourier coefficient of the Boolean function f, calculated on character χ_S
F	The accepting states of a DFA so that $F \subseteq Q$; additionally, in PAC model, it denotes a target concept class: $F = \cup_{n \geq 1} F_n$
F_n	A target concept over the instance space C_n
h	A hypothesis
H	A hypothesis class
$I(f)$	The average sensitivity of a Boolean function f
$\mathrm{Inf}_i(f)$	The influence of a variable i on the Boolean function f
K	A constant number
L	A decision list
$L(A)$	The set of strings accepted by a DFA A
m	An integer value corresponding to the maximum of the statistically relevant delay values inside the stages of an Arbiter PUF; moreover, it denotes the number of examples
M	An integer value corresponding to the maximum of the statistically relevant delay values in an Arbiter PUF
n	Number of bits, or Boolean variables corresponding to a challenge

$\mathbb{N}$	The set of all natural numbers
N	The number of ring oscillator in an RO-PUF
N_{mis}	The upper bound of the mistakes of the Perceptron algorithm
$O(\cdot)$	Big O notation
p	Some probability
$p(\cdot)$	Some polynomial
P_n	A Perceptron with n inputs and single output
$\mathcal{P}$	The hyperplane bounding the half-spaces S^0 and S^1
q	A state in a DFA
q_0	The initial state of a DFA
Q	The set of states of a DFA
R	The maximum Euclidean length of a vector, e.g., $\boldsymbol{\Phi}$
S	Some set such that $S \subseteq \{1,\ldots,n\}$; moreover, in PAC model, it denotes a set of examples
S^0 and S^1	Two half-spaces; the sets containing positive and negative examples, respectively.
$\mathbf{u}$	The solution vector with $\|\mathbf{u}\| = 1$
u_i	Parity of challenge bits
U	A set of challenge-response pairs
υ	A value in the set $\{0,1\}$
V_n	The set of Boolean attributes or variables
y	A response of a PUF
$\mathcal{Y}$	The set of the responses of a PUF

Operators

$\boldsymbol{\Phi} \cdot \omega$	The inner product of vectors, e.g., ω and $\boldsymbol{\Phi}$
$\neg$	The complement operator
$\|\cdot\|$	The Euclidean length of a vector
$!$	Factorial
$\in$	Element of a set
$\in_D$	Chosen from a set according to a distribution
$\ni$	$S \ni i$ is equivalent to $i \in S$
$\subseteq$	A subset of, equals to
$\cap$	Set intersection
$\oplus$	Logical exclusive-or (XOR)
$\otimes$	Tensor product
$\vee$	Logical OR
$\wedge$	Logical AND
$*$	Convolution operator
$|\cdot|$	The cardinality operator; moreover, the length of a string
$\lceil \cdot \rceil$	Rounds a number to the nearest integers greater than or equal to that

$\mathbf{E}_{c \in \mathcal{U}}[\cdot]$ The expectation over uniformly chosen random examples c

$\ln$ Natural logarithm

$\log_2$ Logarithm base 2

$VC_{dim}(\cdot)$ Vapnik–Chervonenkis dimension

$size(\cdot)$ The mapping that associates a natural number $size(f)$ with a target concept $f \in F$

$\mathrm{sign}(z)$ For a real-valued number $z \in \mathbb{R}$, $\mathrm{sign}(z) = 1$, if $z \geq 0$; otherwise, $\mathrm{sign}(z) = -1$

List of Figures

Fig. 1.1   A simplified IC supply chain   3

Fig. 1.2   PUFs, as an RoT   4

Fig. 1.3   Our general roadmap for proving the learnability of known, and widely used families of PUFs   6

Fig. 2.1   An example of a DL.   12

Fig. 2.2   The schematic of a Perceptron   15

Fig. 3.1   Schematic of an Arbiter-PUF   24

Fig. 3.2   The probability distribution of the delays in an Arbiter PUF   25

Fig. 3.3   A DFA representing an Arbiter PUF   27

Fig. 3.4   The shrunk DFA representing an Arbiter PUF   28

Fig. 3.5   Block diagram of the PAC learning algorithm   29

Fig. 4.1   Schematic of an XOR Arbiter PUF   38

Fig. 4.2   Schematic of an Arbiter PUF   39

Fig. 4.3   Block diagram of the PAC learning framework applied to learn an XOR Arbiter PUF.   40

Fig. 4.4   Upper bound of the number of CRPs required to PAC learn an XOR Arbiter PUF   42

Fig. 5.1   An RO-PUF with N ring-oscillators   52

Fig. 5.2   An example of the upper bound of the number of CRPs required to PAC learn an RO-PUF.   55

Fig. 5.3   Results of launching our PAC learning attack against RO-PUFs   56

Fig. 6.1   The schematic of a BR-PUF with n stages   61

Fig. 6.2   The schematic of a TBR-PUF with n stages.   62

Fig. 6.3   The roadmap of our framework for PAC learning of BR-PUF family   66

Fig. 6.4   The relation between the theoretical upper bound on the error of the final model returned by Adaboost   69

List of Tables

Table 6.1   Statistical analysis of the 2048 CRPs,
            given to a 64-bit BR-PUF [121]......................... 63
Table 6.2   Statistical analysis of the 30000 CRPs,
            given to a 64-bit BR-PUF 64
Table 6.3   The average sensitivity of n-bit BR-PUFs and the number
            of CRPs collected to compute that...................... 65
Table 6.4   The results of K-junta test on a BR-PUF.................. 68
Table 6.5   Experimental results for learning 64-bit BR-PUF
            and TBR-PUF, when $m = 15$........................... 70
Table 6.6   Experimental results for $m = 100$ (the same setting
            as for the Table 6.5)................................. 71
Table 6.7   Experimental results for $m = 1000$ (the same setting
            as for the Tables 6.5 and 6.6).......................... 71

List of Algorithms

Algorithm 1   The skeleton of a Boosting algorithms.   19
Algorithm 2   PAC learning algorithm A for f_{PUF}.   31
Algorithm 3   The algorithm for PAC learning of a k-DL as proposed
               by [90].   53

Abstract

Along with the expansion of the application of integrated circuits (ICs), the risk of security vulnerabilities of them has been rising dramatically. Diverse types of successful attacks against ICs have been launched by adversaries taking advantage of those vulnerabilities to expose confidential information, e.g., intellectual property (IP), and customer data. As a remedy for the shortcomings of traditional security measures taken to protect ICs, PUFs appear to be promising candidates that are intended to offer instance-specific functionality. While cryptographic mechanisms enjoying this privilege have been emerging, the vulnerability of PUFs to different types of attacks has been demonstrated. Among these attacks, a great deal of attention has been paid to machine learning (ML) attacks, aiming at modeling the challenge-response behavior of PUFs. So far the success of ML attacks has relied on trial and error, and consequently, ad hoc attacks and their corresponding countermeasures have been developed.

This thesis aims to address this issue by providing the mathematical proofs of the vulnerability of various PUF families, including Arbiter, XOR Arbiter, ring oscillator, and bistable ring PUFs, to ML attacks. To achieve this goal, for the assessment of these PUFs a generic framework is developed that includes two main approaches. First, with regard to the inherent physical characteristics of the PUFs mentioned above, fit-for-purpose mathematical representations of them are established, which adequately reflect the physical behavior of those primitives. To this end, notions and formalizations, being already familiar to the ML theory world, are reintroduced in order to give a better understanding of why, how, and to what extent ML attacks against PUFs can be feasible in practice. Second, polynomial time ML algorithms are explored, which can learn the PUFs under the appropriate representation. More importantly, in contrast to previous ML approaches, not only the accuracy of the model mimicking the behavior of the PUF but also the delivery of such a model is ensured by our framework.

Besides off-the-shelf ML algorithms, we apply a set of algorithms originated in property testing field of study that can support the evaluation of the security of PUFs. They serve as a "toolbox", from which PUF designers and manufacturers can choose the indicators being relevant to their requirements. Last but not least, on the basis of learning theory concepts, this thesis explicitly states that the PUF families studied here cannot be considered as an ultimate solution to the problem of insecure ICs. Furthermore, we believe that this thesis can provide an insight into not only the academic research but also the design and manufacturing of PUFs.

Chapter 1
Introduction

> *Unfortunately, none of the candidate [PUF] constructions have a proof of computational security, and further, most, if not all, of them have been shown to be susceptible to ML attacks. [In this context] Gassend et al. [38] write: An important direction of research is to find a circuit that are provably hard to break [...]*
>
> [49]

1.1 Motivation

Among the characteristics of today's integrated circuits (ICs) are the wide range of applications and the increasingly interconnected nature, which highlight the importance of the security of ICs. Electronic components used in household appliances and the aerospace industry can be within two ends of the wide spectrum of IC applications. Within this broad range of applications, the internet of things (IoT) with a diverse set of connected devices and objects has become increasingly more popular in recent years. While such technological advances continue to gather pace, their security still lags far behind.

Due to attacks against ICs, as building blocks of the connected devices, the risk of exposing confidential information, e.g., intellectual property (IP) and customer data, as well as endangering the brand image of a manufacturer arises [1]. This imposes serious challenges to IC manufacturers and strictly enforce them to redefine their design goals by adding the security to the former goals, namely cost, power consumption, performance, and reliability [92]. In this regard, the natural and important questions would be (1) to what extent sacrificing the security of ICs can influence the security of the system embodying them, and (2) how the security of ICs has been considered by IC designers and manufacturers in the past and the present. In the

© Springer International Publishing AG 2018
F. Ganji, *On the Learnability of Physically Unclonable Functions*, T-Labs
Series in Telecommunication Services, https://doi.org/10.1007/978-3-319-76717-8_1

following sections, these questions are answered by underlining the high relevance of IC security as well as briefly introducing attacks against ICs and the corresponding countermeasures.

1.1.1   Hardware Root of Trust

In order to answer the questions mentioned above, this section aims at stressing the importance of securing ICs, upon which a root-of-trust (RoT) can be built, and how this issue has been addressed so far. We begin with the following definition of RoT.

Definition 1.1.1 (*RoT* [116]) It is an element of a system that offers services to verify the achievement of security-related goals, e.g., system and data integrity (i.e., data being unaltered) and confidentiality (i.e., data access restricted to authorized entities).

Following Definition 1.1.1 as well as the specifications provided by standardization bodies (e.g., National Institute of Standard and Technology, NIST), an RoT is a primitive that consists of hardware and/or software offering trusted, security-critical functions [19]. Furthermore, as its name implies, an RoT establishes a chain of trust ensuring the owner of the device about the security of that. Due to the immutability, a smaller attack surface, and a more reliable behavior, hardware RoTs are preferred options for providing services defined in Definition 1.1.1 [19, 21, 116]. We stress that since levels of design abstraction (i.e., protocols, software, etc.) and the methodology to establish an RoT model are beyond the scope of this thesis, we solely focus on the hardware RoT (for more details cf. [119]).

Although several requirements that must be fulfilled by a hardware RoT have been formulated in the literature, here we stick to the following one, which is the most relevant to our work. A hardware RoT must provide a mechanism for generating and storing a key, which can be altered or revealed (e.g., by reverse-engineering) either in prohibitively expensive or practically impossible fashion [60]. The hardware RoT orthodoxy is a secret key embedded in the hardware [87], for instance, a key stored in the non-volatile memories of the IC. Unfortunately, the vulnerability of such legacy key storage methods to semi- and fully-invasive attacks, has been reported in the literature [47, 58, 112, 115]. It has been demonstrated that even in the presence of sophisticated countermeasures, it is hard to stop an adversary attempting to circumvent these security measures and gain access to the memories. Accordingly, in the absence of an RoT, the chain of trust cannot be established, and the security of the whole system can be compromised.

1.1.2   Fragile Security of ICs

As is evident from the above discussion, IC designers and manufacturers have to come up with security policies and measures to protect the hardware RoT [119]. In addition, several successful attacks, ranging from a simple to an orchestrated one,

Fig. 1.1 A simplified IC supply chain (inspired by [21, 67])

have been mounted against ICs, which particularly take advantage of the lack of secure hardware RoT. Here we briefly consider such attacks against ICs, which are the most relevant to the scope of this thesis, and their roots.

The problem of IC insecurity has been escalating due to the horizontal semiconductor business model, in which the IC fabrication is mainly outsourced to a contract foundry (usually) located in a foreign country [59]. Figure 1.1 illustrates a model of a simplified IC supply chain, which can provide a better understanding of the goal and the harmfulness of attacks against ICs. The chain begins with designing and procuring IP designs. After the verification phase, an IC is manufactured and tested, and fault-free ICs are packaged. It is important to emphasize that in the system integration phase, fake and low-quality components may enter the chain as well [92]. The message conveyed here is that during all phases, when the IC is designed and produced, various hardware-based attacks can be mounted against an IC. In this thesis, our primary focus lies on IP piracy and overbuilding attacks (see [92] for further details on other types of attacks).

IP piracy and overbuilding attacks refer to cases, where a user and/or a foundry, who is given access to the IP, misuse, and pirate the IP to build ICs more than the desired number, ordered by the IP owner or the rights holder [92]. The goals of this attack can be stealing the IP design and identifying the trade secret. In another case, the original design and/or component, e.g., the IC, can be forged or replicated by an adversary, with an aim to eavesdrop the sensitive information, modify the IC functionality, mount the denial of service attacks, and reduce the reliability of the IC [113]. These attacks can be launched in virtually all phases of the IC supply chain [92].

In addition to the huge cost and losses (e.g., loss of image) for IC manufacturer, such attacks can affect an end-user, ranging from government, industry, and business to ordinary consumers. To alleviate or prevent these adverse effects, mechanisms should be employed to distinguish between an overbuilt IC and a real one. These mechanisms should include not only embedding a hardware RoT in ICs but also protecting such RoT from attacks, e.g., those introduced previously. To address this, and as a countermeasure against these attacks, Gassend et al. were the first to suggest applying a silicon-unique RoT [38], called a physically unclonable function (PUF). In this respect, it is claimed that a PUF can provide IC-unique authentication as well as key generation.

1.2 Physically Unclonable Functions

The invention of PUFs was a milestone in the development of device authentication and key generation methods, as explained before. The basic idea behind the concept

Fig. 1.2 PUFs, as an RoT, must provide trusted hardware resources (i.e., chip-unique key generation and key storage). These resources serve a ground for developing cryptographic primitives (inspired by [21, 67])

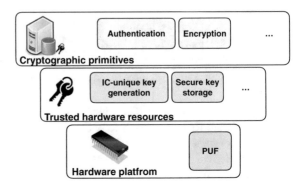

of PUFs is to take advantage of physical characteristics of the IC corresponding to manufacturing process variations, which make each IC slightly different from others. Regarding theses instance-specific, and inherent physical properties of the PUFs, they have been nominated as a hardware RoT. More specifically, the IC-unique key generation and storage, so-called trusted hardware resources, are provided by a PUF embedded in a hardware platform, i.e., the IC [21]. These resources can be employed to support the cryptographic primitives, see Fig. 1.2.

For PUFs, due to the manufacturing process variations, it is possible to generate virtually unique *responses* as outputs, when the instance is given some inputs called *challenges*. Therefore, PUFs can be utilized as either device fingerprints for secure authentication or as a source of entropy in secure key generation scenarios. In this case, there is no need for permanent key storage, since the desired key is generated instantly upon powering up the device. Due to the physical properties of the PUFs, they are assumed to be *unclonable* and *unpredictable*, and therefore trustworthy and robust against attacks [67].

Regarding how the physical characteristics of the IC embodying a PUF are exploited, several PUF instances have been introduced in recent years [67], e.g., Arbiter, XOR Arbiter, ring oscillator (RO-), and bistable ring (BR-) PUFs. While the PUF manufacturers have been contributing to improve the design, and consequently, the security of the PUFs, adversaries are simultaneously developing non-invasive and semi-invasive attacks against these primitives [46, 47, 95, 109]. For instance, it has been shown that an Arbiter PUF can be fully characterized by conducting semi-invasive temporal photonic emission analysis [109]. As another example of semi-invasive attacks, it has been stated that RO-PUFs are subject to semi-invasive electromagnetic (EM) side channel analysis [78].

On the other hand, machine learning (ML) attacks are one of the most common types of non-invasive attacks against PUFs, whose popularity stems from their characteristics, namely being cost-effective and non-destructive. From the very beginning of the era of PUFs, designers were doubtful, whether the input-output (i.e., so-called *challenge-response*) behavior of a PUF can be learned cf. [38]. This question is relevant since, in these attack scenarios, the adversary *solely* observes the challenge-response behavior of the targeted PUF by applying a relatively small set of challenges to a PUF and collects the responses to those challenges. Afterwards, by employing ML techniques, the adversary can build a model of the challenge-response behavior of the PUF, which can *predict* the responses of the PUF to new, arbitrarily chosen challenges.

1.3 Thesis Statement

As explained above, in ML attack scenarios, a relatively small subset of challenges along with their respective responses is collected by the adversary, attempting to come up with a model describing the challenge-response behavior of the PUF. So far the success of existing modeling attacks, so-called *empirical* ML attacks, relies on simple trial-and-error approaches cf. [83, 95, 97, 98]. The most serious shortcoming of these ML approaches is that after the learning phase, the delivery of a model describing the challenge-response behavior of the PUF is not ensured. Furthermore, the existing models of the challenge-response behavior of the PUFs fail to reflect the physical properties of these primitives in some attack scenarios. This failure results in applying less powerful adversarial models. These have led the designers to conjecture that PUFs can be proved to be computationally secure after all [49].

1.3.1 Problem Statement

As stated above, in the absence of a detailed analysis and mathematical proofs of the security of PUFs, ad hoc attacks and their respective countermeasures have been developed. We aim to address this issue by providing *mathematical proofs* of the vulnerability of different PUF families (including Arbiter, XOR Arbiter, RO-, and BR-PUFs) to ML attacks. In order to develop such mathematical framework for the assessment of the security of PUFs, we take the following necessary steps:

- Establishing fit-for-purpose mathematical representations of different PUFs. To this end, we should accurately capture the physical characteristics of a PUF, and translate it to mathematical representations of them.
- Developing approaches to assess the security of PUFs under the proposed representation. When applying these methods, the delivery of a model describing the challenge-response behavior of the PUF must be ensured. Furthermore, it is desired to compute the maximum number of examples that should be collected from the PUF before launching the ML attack.

In a nutshell, by providing mathematical proofs, this thesis attempts to answer the following question. Are PUF families, studied by us in this thesis, provably suitable solutions to the problems concerning the security of ICs?

1.3.2   Our Attack Model

Here, we informally introduce our attack model, for further details on that see Chap. 2. Aiming at mathematically *cloning* known families of PUFs, namely, Arbiter, XOR Arbiter, RO-, and BR-PUFs, we collect a set of challenge-response pairs (CRPs) to provide a model that can *approximately* predict the response of the PUF to an arbitrarily chosen challenge. In our framework, the accuracy of the model delivered by the machine, and the probability of delivering such accurate model (so-called confidence level) can be defined beforehand. Moreover, the maximum number of examples required to run our ML algorithms is known before the learning phase. These are in contrast to empirical ML attacks that have been proposed in the literature so far.

As depicted in Fig. 1.3, we take the first step by exploring the physical, inherent properties of a PUF. Additionally, in some scenarios, the mathematical model of the internal functionality of the PUF is known (e.g., Arbiter PUFs, see Chap. 3), which can be helpful to build a proper representation of the PUF. Afterwards, we apply an algorithm to learn the PUF for prescribed levels of accuracy and confidence under the representation, established in previous steps. More formally, for a family of PUF, we

- take advantage of the physical, inherent characteristics and special features of the PUF family, which are helpful to establish polynomial-sized representations of the PUFs, and
- provide polynomial time algorithms that can generate such polynomial-sized representation of the PUF, when being fed with a set of CRPs.

Finally, we evaluate the feasibility and effectiveness of our framework applied to real-world PUFs and/or compare our theoretical findings with the practical results reported in the literature.

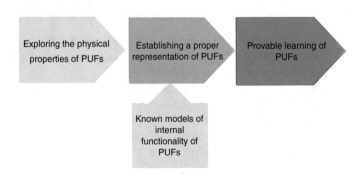

Fig. 1.3 Our general roadmap for proving the learnability of known, and widely used families of PUFs (i.e., Arbiter, XOR Arbiter, RO-, and BR-PUFs)

1.3.3 Thesis Contributions

In this thesis, we present provable ML attacks against several families of PUFs, namely, Arbiter, XOR Arbiter, RO-, and BR-PUFs. Over the years, when we have been writing this thesis, various empirical ML attacks have been proposed in the literature. In contrast to these approaches, we have proved that PUFs, studied in this thesis, cannot be considered as "circuits that are provably hard to break".

First, we have considered the family of Arbiter PUFs that is widely accepted and studied in the hardware security community. The principle behind the design of these PUFs is that the delay differences between symmetrically designed electrical paths on an IC can be utilized to generate a response, when the IC is fed by a challenge. We have demonstrated that under a well-established representation, these PUFs can be learned for given levels of accuracy and confidence [36]. In our study, the impact of the limited range of the electrical path delays on the learnability of the Arbiter PUFs has been extensively investigated and has its novelty value.

Secondly, we have studied XOR Arbiter PUFs as a modified structure of Arbiter PUFs, in which non-linear effects are added to the PUF in order to impair the effectiveness of ML attacks. At the beginning of the XOR Arbiter PUFs era, it was assumed that when XORing a number of Arbiter chains to generate the response of the PUF, the PUF would be more robust against ML attacks. We have established a theoretical upper bound on the number of arbiter chains of an XOR Arbiter PUF, beyond which the PUF cannot be characterized in polynomial time by applying pure ML methods [35]. We have further elaborated on how the technological constraints of existing ICs can pose difficulties for manufacturers that implement an XOR Arbiter PUF, whose number of arbiter chains exceeds the theoretical upper bound. Our results have been acknowledged, and provide a firm basis for not only performing further research but also designing new PUF architectures [89, 123, 124].

Thirdly, in our study on RO-PUFs, we have demonstrated that inherent characteristics of RO-PUFs, although being *hidden* from an adversary, contribute to the success of heuristic-based attacks and our framework as well. The most substantial contribution of our study is establishing a fit-for-purpose representation of RO-PUFs that can be easily and rapidly built by collecting CRPs from PUFs [34].

Last but not least, we have provided a provable framework for ML attacks against a PUF family, whose underlying mathematical model is unknown, namely BR-PUFs [32]. So far virtually all ML attacks have been relying on the assumption that a mathematical model of the PUF functionality is known. While this may not be true in some cases, attention should be paid to this important aspect of ML attacks. When we were developing the framework mentioned above, we observed an interesting phenomenon that has also been noticed by several papers on experimental research. However, none of them has correctly and precisely pinpointed the mathematical origin of that. We have proved that when applying a challenge to a PUF, the response the PUF is not determined equally by all challenge bits, but a subset of the bit positions. In addition, to the best of knowledge, we – for the first time – endeavor to apply the Boolean and Fourier analyses in order to assess the security of PUFs. In this regard, we have introduced new metrics and notions, e.g., *average sensitivity of Boolean functions*, which are well-known and widely used in modern cryptography and may provide special insights into the physical design of secure PUFs in the future. The paper covering the results of the study on BR-PUFs has been well received by the hardware security community and invited to the Journal of Cryptographic Engineering as one of "the best rated papers" [40]. The extension of the paper [32] published by this journal introduces further notions from property testing field of study and has been noticed as "providing an important addition to the toolbox of PUF designers" [40].

Finally, we believe that this thesis can help to fill the gap between ML theory and hardware security, in particular, by contributing to the design, and the assessment of the security of PUFs in the real world. More specifically, this thesis demonstrates that provable algorithms and approaches that are common knowledge in the ML theory should be taken into account, when designing PUFs and assessing the security of them.

1.4   Outline of the Thesis

Chapter 2 presents the background information and notations required to understand the general concept of PUFs and our attacks. Chapter 3 describes how an Arbiter PUF can be subject to our provable learning attack. In Chap. 4, our ML attack against XOR Arbiter PUFs is discussed. Chapter 5 demonstrates that the security of RO-PUFs can be broken by launching our attack. The aim of Chap. 6 is to state that BR-PUFs – even without knowing a model representing their internal functionality – are vulnerable to our ML attack. Chapter 7 gives a brief introduction to work derived from and built upon the results of our learning framework. Finally, Chap. 8 draws the main conclusions from this thesis and discusses directions for future work.

Chapter 2
Definitions and Preliminaries

Check for
updates

This chapter focuses on the background information and notations required to understand the general concept of PUFs, PAC model and several mathematical representations established to launch ML attacks against PUFs. The material collected and
reworked in this chapter has been first presented in [32–36].

Overview of this chapter: In addition to introducing the notations used in this
thesis (Sect. 2.1), this chapter is devoted to the background information related to the
notion of PUFs (Sect. 2.2), basics of Boolean analysis (Sect. 2.3), and our ML model
(Sect. 2.6). Moreover, one of our main results has been formulated in Sect. 2.3.3 that
is Theorem 2.3.2.

2.1 Notations

In this thesis, vectors are denoted by a bold character, e.g., $\boldsymbol{c}$, whose elements are
indexed by an index $i \geq 1$ between brackets, for instance, $\boldsymbol{c}[1]$. A boolean string is
denoted by a normal character, e.g., c, where its elements are $c = c_1 c_2 \ldots c_n$. The
concatenation of two strings, e.g., c and c', is shown as cc'. A learning algorithm is
displayed in `teletype font`, e.g., an algorithm A. Random variables are shown
in the calligraphic font, e.g., $\mathcal{A}$. A list of the operators used in this thesis is provided
in the list of symbols on p.xxx.

2.2 PUFs

Note that elaborate and formal definitions as well as formalizations of PUFs are
beyond the scope of this work, and for more details, the reader is referred to
[8, 9]. In general, PUFs are physical input to output mappings, which map given

© Springer International Publishing AG 2018

F. Ganji, *On the Learnability of Physically Unclonable Functions*, T-Labs
Series in Telecommunication Services, https://doi.org/10.1007/978-3-319-76717-8_2

challenges to *responses*. Intrinsic properties of the physical primitive embodying the PUF determine the characteristics of this mapping. An *ideal* PUF exhibits the following security-related properties.

Definition 2.2.1 For a given random instance of a PUF, i.e., $f_{PUF}(\cdot)$, let it be defined by the mapping $f_{PUF} : C \rightarrow \mathcal{Y}$, where $f_{PUF}(c) = y$. An ideal f_{PUF} exhibits the following properties.

1. Evaluable: in polynomial time we can evaluate f_{PUF}.
2. Unique: the mapping f_{PUF} is instance-specific.
3. Reproducible: applying same challenges to f_{PUF} results in *close* responses with respect to a chosen distance metric.
4. Unclonable: it is (almost) impossible to construct another mapping (i.e., physical entity) g_{PUF} so that $g_{PUF} \approx f_{PUF}$.
5. Unpredictable: for a given set $U = \{(c_i, y_i) \mid y_i = f_{PUF}(c_i)\}$, it is (almost) impossible to predict a response $y_r = f_{PUF}(c_r)$, where c_r is a random challenge and $(c_r, y_r) \notin U$.
6. One-way: by applying a challenge c, drawn from a uniform distribution on $\{0, 1\}^n$, we obtain $y = f_{PUF}(c)$ so that

$$\Pr[\text{A}(f_{PUF}(c)) = c] < 1/p(n),$$

where $p(\cdot)$ is any positive polynomial. This means that the probability that any probabilistic polynomial time algorithm or physical procedure A can output c is negligible, cf. [94]. In other words, it is *hard* to find the challenge c, if the respective response of a random instance of the PUF family is known, and the adversary can evaluate the PUF only a polynomial number of times [67].

Regarding the number of possible challenges, two main classes of PUFs, namely strong PUFs and weak PUFs have been discussed in the literature [42]. A PUF with an exponential challenge space is called a *strong PUF*, whereas a *weak PUF* does not fulfill this requirement (for a formal definition see [42, 94]). In this section we do not limit our focus to a specific class of PUF, however, we elaborate on this issue in next chapters.

Last but not least, without limiting the generality of our approach, we assume that CRPs collected from a PUF is noiseless (unless it is explicitly stated, e.g., in Sect. 4.4). The term *noise* in the PUF-related literature refers to the observation that applying the same challenge may result in obtaining different responses due to the environmental changes, see, e.g. [67]. This issue must have been resolved by the manufacturer, and is beyond the scope of this thesis. For more details, the reader is referred to numerous mechanisms proposed in the literature, e.g., [70, 71].

2.3 Boolean Functions

Defining PUFs as mappings (see Sect. 2.2), the most natural mathematical model for them are Boolean functions over the finite field $\mathbb{F}_2$. Note that Boolean functions considered in this thesis can be seen solely as an *approximation* of a mapping associated with a physical primitive, namely a PUF as defined by Definition 2.2.1.

Let $V_n = \{c_1, c_2, \ldots, c_n\}$ denote the set of Boolean attributes or variables, where each attribute can be *true* or *false*, commonly denoted by "1" and "0", respectively. In addition, $C_n = \{0, 1\}^n$ contains all binary strings with n bits. We associate each Boolean attribute c_i with two *literals*, i.e., c_i, and $\neg c_i$ that is complement of c_i. An *assignment* is a mapping from V_n to $\{0, 1\}$, i.e., the mapping from each Boolean attribute to either "0" or "1". In other words, an assignment is an n-bits string, where the ith bit of this string indicates the value of c_i (i.e., "0" or "1").

An assignment is mapped by a Boolean formula into the set $\{0, 1\}$. Thus, each Boolean attribute can also be thought of as a formula, i.e., c_i and $\neg c_i$ are two possible formulas. If by evaluating a Boolean formula under an assignment we obtain "1", it is called a *positive example* of the concept represented by the formula or otherwise a *negative example*. Each Boolean formula defines a respective Boolean function $f : C_n \rightarrow \{0, 1\}$. The conjunction of Boolean attributes (i.e., a Boolean formula) is called a *term*, and it can be true or false ("1" or "0") depending on the value of its Boolean attributes. Similarly, a *clause* that is the disjunction of Boolean attributes can be defined. The number of literals forming a term or a clause is called its size. The size 0 is associated with only the term **true**, and the clause **false**.

In the related literature several representations of Boolean functions have been introduced, e.g., juntas, Monomials (M_n), Decision Trees (DTs), and Decision Lists (DLs), cf. [84, 90].

Definition 2.3.1 A Boolean function depending on solely an unknown set of at most k variables is called a k-junta. Formally, there exist k indices $1 \leq i_1 < i_2 < \cdots < i_k \leq n$ and a function $h : \{0, 1\}^k \rightarrow \{0, 1\}$ such that the following holds [103]:

$$\forall c \in \{0, 1\}^n : \ f(c) = h(c_{i_1}, \ldots, c_{i_k}).$$

Definition 2.3.2 A monomial $M_{n,k}$ defined over V_n is the conjunction of at most k clauses each having only one literal.

Definition 2.3.3 A DT is a binary tree, whose internal nodes are labeled with a Boolean variable, and each leaf with either "1" or "0". A DT can be built from a Boolean function in this way: for each assignment a unique path form the root to a leaf should be defined. At each internal node, e.g, at the ith level of the tree, depending on the value of the ith literal, the labeled edge is chosen. The leaf is labeled with the value of the function, given the respective assignment as the input. The *depth* of a DT is the maximum length of the paths from the root to the leafs. The set of Boolean functions represented by DTs of depth at most k is denoted by k-DT.

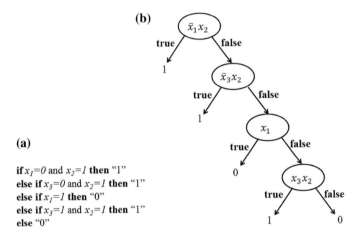

Fig. 2.1 a A sample of "if- then- else if-...- else" rules. **b** Diagram of the DL corresponding to this rule [34]

Definition 2.3.4 A DL is a list L that contains r pairs $(f_1, v_1), \ldots, (f_r, v_r)$, where the Boolean formula f_i is a term and $v_i \in \{0, 1\}$ with $1 \leq i \leq r - 1$. For $i = r$, the formula f_r is the constant function $v_r = 1$. A Boolean function can be transformed into a DL, where for a string $c \in C_n$ we have $L(c) = v_j$, and j is the smallest index in L so that $f_j(c) = 1$. This relationship implies that common patterns are listed at the top of the DL, whereas the exceptions can be found at the end of that. k-DL denotes the set of all DLs, where each f_i is a term of maximum size k.

A useful interpretation of DLs is that they define a concept, which follows a general pattern but with some exceptions, or alternatively, it can be thought of being an extended "if- then- else if-...- else" rule (see Fig. 2.1).

2.3.1 Linearity of Boolean Functions

Here, our focus is on *Boolean linearity*, which must not be confused with the linearity over other domains, different from $\mathbb{F}_2$. A linear Boolean function $f : \{0, 1\}^n \to \{0, 1\}$ features the following equivalent properties, cf. [84]:

- $\forall c, c' \in \{0, 1\}^n : f(c + c') = f(c) + f(c')$
- $\exists a \in \{0, 1\}^n : f(c) = a \cdot c$.

Equivalently, we can define a linear Boolean function f as follows. There is some set $S \subseteq \{1, \ldots, n\}$ such that $f(c) = f(c_1, c_2, \ldots, c_n) = \sum_{i \in S} c_i$.

Boolean linearity or linearity over $\mathbb{F}_2$ is closely related to the notion of correlation immunity. A Boolean function f is called k-**correlation immune**, if for any assignment c chosen randomly from $\{0, 1\}^n$ it holds that $f(c)$ is independent of any k-tuple

$(c_{i_1}, c_{i_1}, \ldots, c_{i_k})$, where $1 \leq i_1 < i_2 < \cdots < i_k \leq n$. Now let $\deg_{\mathbb{F}_2}(f)$ denote the degree of the $\mathbb{F}_2$-polynomial representation of the Boolean function f. It is straightforward to show that such representation exists. Siegenthaler proved the following theorem, which states how correlation immunity can be related to the degree of f.

Theorem 2.3.1 (Siegenthaler Theorem[84, 106]) *Let* $f : \{0, 1\}^n \rightarrow \{0, 1\}$ *be a Boolean function, which is k-correlation immune, then* $\deg_{\mathbb{F}_2}(f) \leq n - k$.

2.3.2 Average Sensitivity of Boolean Functions

The Fourier expansion of Boolean functions serves as an excellent tool for analyzing them, cf. [84]. In order to define the Fourier expansion of a Boolean function $f : \mathbb{F}_2^n \rightarrow \mathbb{F}_2$ we should first define an encoding scheme as follows. $\chi(0_{\mathbb{F}_2}) := +1$, and $\chi(1_{\mathbb{F}_2}) := -1$. Now the Fourier expansion of a Boolean function can be written as

$$f(c) = \sum_{S \subseteq [n]} \hat{f}(S) \chi_S(c),$$

where $[n] := \{1, \ldots, n\}$, $\chi_S(c) := \prod_{i \in S} c_i$, and $\hat{f}(S) := E_{c \in \mathcal{U}}[f(c)\chi_S(c)]$. Here, $E_{c \in \mathcal{U}}[\cdot]$ denotes the expectation over uniformly chosen random examples. In addition, the degree of the Fourier expansion of a function f, $\deg(f)$, is

$$\deg(f) := \max\{|S| : \hat{f}(S) \neq 0\}.$$

Moreover, the **influence of variable** i on $f : \mathbb{F}_2^n \rightarrow \mathbb{F}_2$ is defined as

$$\mathrm{Inf}_i(f) := \mathrm{Pr}_{c \in \mathcal{U}}[f(c) \neq f(c^{\oplus i})],$$

where $c^{\oplus i}$ is obtained by flipping the i-th bit of c. Note that $\mathrm{Inf}_i(f) = \sum_{S \ni i}(\hat{f}(S))^2$, cf. [84]. Next we define the **average sensitivity** of a Boolean function f as

$$\mathrm{I}(f) := \sum_{i=1}^{n} \mathrm{Inf}_i(f).$$

The notion of average sensitivity was first introduced by Kahn et al. [52], and has been widely applied to study the properties of Boolean functions (for a survey on relevant results and their ubiquitous applications see [84]). As a prime example, it is a known result and common knowledge that randomly chosen n-bit Boolean functions have an expected average sensitivity of exactly $n/2$, cf. [84]. This means that if the output of a random function f is ± 1 with probability $1/2$, irrespective of the inputs, the expected average sensitivity of the function f is $n/2$.

2.3.3 Non-linearity of PUFs over $\mathbb{F}_2$ and the Existence of Influential Bits

Section 2.3.1 introduced the notion of Boolean linearity. Focusing on this notion and taking into account the definition of PUFs mentioned in Sect. 2.2, now we prove the following theorem that is our first important result. For all PUFs, when represented as a Boolean function, it holds that their degree as $\mathbb{F}_2$-polynomial is strictly greater than one.

Theorem 2.3.2 *For every strong PUF* $f_{PUF} : \mathbb{F}_2^n \to \mathbb{F}_2$, *we have* $\deg_{\mathbb{F}_2}(f_{PUF}) \geq 2$.

Proof Towards contradiction assume that the Boolean function f_{PUF}, which fulfills the requirements mentioned by Definition 1, is linear over $\mathbb{F}_2$. From the unpredictability of f_{PUF} it follows that the adversary has access to a set of CRPs $U = \{(c, y) \mid y = f_{\text{PUF}}(c) \text{ and } c \in \mathcal{C}\}$, which are chosen uniformly at random, however, the adversary has only a negligible probability of success to predict a new random challenge $(c', \cdot) \notin U$ (as he cannot apply f_{PUF} to this unseen challenge). Note that the size of U is actually polynomial in n. Now, by the definition of linearity over $\mathbb{F}_2$, cf. Sect. 2.3.1, we deduce that the only linear functions over $\mathbb{F}_2$ are the Parity functions, see also [84, 106]. However, there are well-known algorithms to PAC learn Parity functions in general [26, 48]. Thus, now we simply feed the right number of samples from our CRP set U into such a PAC learner. For the right parameter setting, the respective PAC algorithm delivers then with high probability an ε-approximator h for our PUF f_{PUF} such that $\Pr[f(c') = h(c')] \geq 1 - \varepsilon$. This means that with high probability, the response to every randomly chosen challenge can be calculated in polynomial time. This is of course a contradiction to the definition of f_{PUF}, being unpredictable. Hence, f_{PUF} cannot be linear over $\mathbb{F}_2$, i.e., we have $\deg_{\mathbb{F}_2}(f_{\text{PUF}}) \geq 2$. ∎

There is a close connection between this property of PUFs and the notion of correlation immunity introduced by Siegenthaler's Theorem. It is known that any parity function on n inputs is $n - 1$-resilient, and $n - 1$-correlation immune (cf. [84]). As a result of the above discussion, at best, a PUF can be approximated by a parity function on $n - 1$ inputs, and consequently, we deduce that a PUF can be represented by (at best) an $n - 2$-correlation immune function.

Although the condition on $\deg_{\mathbb{F}_2}(f_{\text{PUF}})$, proved by Theorem 2.3.2, is important from our point of view, it cannot specify how bits can influence the responses of a PUF[1]. In fact, in this regard we refer to what has been proved by Kahn et al.; for a Boolean function on n variables that equals "1" with probability p ($p \leq 1/2$), there exists at least one variable with the influence $\Omega(p \log_2(n)/n)$, cf. [52]. This result assures that there exists at least one variable with the influence $\Omega(p \log_2(n)/n)$, although it neither quantifies a certain number of influential bits nor approximates

---

[1]Unlike the degree of the Fourier expansion of a function f ($\deg(f)$), from $\deg_{\mathbb{F}_2}(f)$ it is not straightforward to infer the number of bit positions influencing the response of a Boolean function, cf. [84]

the average sensitivity of a Boolean function. The key message conveyed by Kahn's theorem is that such an influential bit position exists. However, in order to approximate the average sensitivity of a Boolean function representing a PUF, and consequently, the number of influential bits of that, Boolean analyses must be conducted (see Sect. 6.3.2). Note that these Boolean analyses aim to determine an upper bound on the number of influential bits to evaluate the feasibility of PAC learning of a PUF family.

Note on Boolean functions representing a PUF: when interpreting the results of Theorem 2.3.2 and Kahn's theorem in the case of PUFs, Boolean functions within the scope of these theorems have to reflect the characteristics of a PUF, see Definition 2.2.1. For instance, symmetric (i.e., the number of "1"s in the input determines the output), and totally random as well as totally deterministic Boolean functions cannot be good candidates to represent a PUF. More specifically, these are Boolean functions that cannot fulfill the requirements of a PUF, namely, unpredictability, and robustness.

2.4 Linear Threshold Functions

In addition to the Boolean functions, linear threshold functions (LTFs) have been used to represent PUFs. In order to define an LTF, we begin with the definition of a Perceptron (i.e., single-layer Perceptron), see Fig. 2.2.

Definition 2.4.1 (cf. [79]) A Perceptron P_n with n inputs and single output is associated with a function $f : \{0, 1\}^{n+1} \to \{-1, 1\}$, where the output is computed by the Perceptron is:

$$f = \begin{cases} 1, & \text{if } \sum_{i=1}^{n} \omega[i]\Phi[i] \geq 0 \\ -1, & \text{otherwise.} \end{cases}$$

In the above equation, the vector $\omega = (\gamma, \omega[1], \omega[2], \ldots, \omega[n])$ contains real-valued weights $\omega[i]$ determining the contribution of each input element $\Phi[i]$ to the output of the Perceptron. Alternatively, we can formulate the above formula as an LTF

Fig. 2.2 A Perceptron first calculates a linear combination of its inputs, which is finally compared to some threshold to generate the output

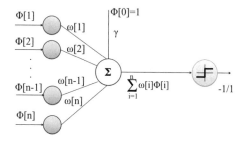

$$f = \text{sign}(\mathbf{\Phi} \cdot \boldsymbol{\omega}).$$

The sets of positive and negative examples of the LTF f are half-spaces S^1 and S^0, where

$$S^1 = \{\mathbf{\Phi} \in \mathbb{R}^n \mid \sum_{i=1}^{n} \omega[i]\mathbf{\Phi}[i] \geq \gamma\},$$

and

$$S^0 = \{\mathbf{\Phi} \in \mathbb{R}^n \mid \sum_{i=1}^{n} \omega[i]\mathbf{\Phi}[i] < \gamma\}.$$

S^1 and S^0 are decision regions of the LTF f that are bounded by the hyperplane $\mathcal{P} : \mathbf{\Phi} \cdot \boldsymbol{\omega} = \gamma$ (for further details see [7]).

Perceptron Algorithm

The Perceptron algorithm is an *online* algorithm invented to learn LTFs efficiently. By online we mean that providing the learner (i.e., learning algorithm) with each example, e.g., $\mathbf{\Phi}_i$, it attempts to predict the response to that example. Afterwards, the actual response (i.e., the label, for instance $f(\mathbf{\Phi}_i)$) is presented to the learner, and then it can improve its hypothesis by means of this information. The learning process continues until all the examples are provided to the learner [65].

Let the input of the Perceptron algorithm be a sequence of r labeled examples $((\mathbf{\Phi}_1, f(\mathbf{\Phi}_1)), \ldots, (\mathbf{\Phi}_r, f(\mathbf{\Phi}_r)))$. The output of the algorithm is the vector $\boldsymbol{\omega}$ classifying the examples. Executing the Perceptron algorithm, it initially begins with $\boldsymbol{\omega}_0 = (\omega_0[1], \omega_0[2], \ldots, \omega_0[n], \gamma) = (0, \ldots, 0)$. When receiving each example (e.g., $\mathbf{\Phi}_j$), the algorithm examines whether $\omega_j[i] \cdot \mathbf{\Phi}_j[i] \geq \gamma_j$ and compares its prediction with the received label. If the label and the prediction of the algorithm differ, ω_j is updated as follows:

$$\omega_{j+1}[k] = \begin{cases} \omega_j[k] - f(\mathbf{\Phi}_j) \cdot \mathbf{\Phi}_j[k] & 1 \leq k \leq n \\ \gamma_j - f(\mathbf{\Phi}_j) & k = n+1. \end{cases}$$

Note that if the prediction and the label of an example agree, no update is performed [104].

Quantifying the performance of an on-line algorithm, the prediction error (i.e., number of mistakes) of the algorithm is taken into account. In this way, the upper bound of the mistakes is defined as a measure of the performance. The Perceptron convergence theorem gives an upper bound of the error that can occur while executing the Perceptron algorithm [29]:

Theorem 2.4.1 (Convergence Theorem of the Perceptron algorithm): *Consider r labeled examples which are fed into the Perceptron algorithm, and $\|\mathbf{\Phi}_i\| \leq R$ ($\|\cdot\|$ denotes the Euclidean length). Let $\boldsymbol{u}$ be the solution vector with $\|\boldsymbol{u}\| = 1$ whose error is denoted by ε ($\varepsilon > 0$). The deviation of each example is defined as*

$d_i = \max\{0, \varepsilon - f(\mathbf{\Phi}_i)(\mathbf{u} \cdot \mathbf{\Phi}_i)\}$, and $D = \sqrt{\sum_{i=1}^{r} d_i^2}$. *The upper bound of the mistakes of the Perceptron algorithm is*

$$N_{mis} = \left(\frac{R+D}{\varepsilon}\right)^2.$$

For the proof, the reader is referred to [29]. Note that when the data is linearly separable, we have $D = 0$ [29]. In order to examine whether the data is linearly separable, we should define a parameter as follows. Let the parameter σ be the minimum distance of any example from $\mathcal{P}$, i.e.,

$$\sigma = \min_{\mathbf{\Phi} \in \Phi} \frac{|\mathbf{\Phi} \cdot \mathbf{\omega}|}{\|\mathbf{\omega}\| \|\mathbf{\Phi}\|},$$

where Φ is the set of all $\mathbf{\Phi}$'s [12]. This equation can be seen as the minimum of the cosine of the angle between the input and the weight vectors. The order of $1/\sigma$ determines whether the data is linearly separable. It has been demonstrated that when $1/\sigma$ is exponential in n, the data is not linearly separable, and consequently, the Perceptron algorithm cannot classify the data [18, 104]. On the other hand, if $1/\sigma$ is polynomial in n, the Perceptron algorithm can be applied.

2.5 Regular Language and Principles of DFAs

We assume that the reader is familiar with regular languages and deterministic finite automata (DFA). Therefore, we only briefly introduce the notations, which will be used throughout this work. We follow the standard notation, as can be found in [4] and [51]. Consider the alphabet $\Sigma = \{0, 1\}$ and the set of all strings Σ^* over Σ. By $|c|$ we denote the length of strings $c \in \Sigma^*$, and by λ the empty string of length $|\lambda| = 0$.

A DFA A is given by $A = (Q, \delta, \Sigma, q_0, F)$ over the alphabet Σ, with Q being the set of states, the initial state q_0, and the accepting states $F \subseteq Q$. The transition function $\delta : Q \times \Sigma \to Q$ is defined as follows. For all $q \in Q, a \in \Sigma$ and $c \in \Sigma^*$ we have $\delta(q, \lambda) = q$ and its canonical continuation to Σ^*, i.e., $\delta(q, ac) = \delta(\delta(q, a), c)$. The set of strings accepted by A is called its accepted language $L(A) := \{c \in \Sigma^* \mid \delta(q_0, c) \in F\}$, i.e., a regular language. A state q_i is *live*, if there exist $c_1, c_2 \in \Sigma^*$ such that $c_1 c_2 \in L(A)$ with $\delta(q_0, c_1) = q_i$ and $\delta(q_i, c_2) \in F$. Otherwise, q_i is called dead.

2.6 Probably Approximately Correct Model

The Probably Approximately Correct (PAC) model provides a firm basis for analyzing the efficiency and effectiveness of ML algorithms. We briefly introduce the model and refer the reader to [56] for more details. In the PAC model the learner, i.e., the learning algorithm, is given a set of *examples* to generate with high probability an approximately correct hypothesis. This can be formally defined as follows. Let $F = \cup_{n \geq 1} F_n$ denote a *target concept class* that is a collection of Boolean functions defined over the *instance space* $C_n = \{0, 1\}^n$. Moreover, according to an arbitrary probability distribution D on the instance space C_n each example is drawn. Assume that a hypothesis $h \in H_n$ is a Boolean function over C_n, it is called an ε-approximator for $f \in F_n$, if

$$\Pr_{c \in_D C_n} [f(c) = h(c)] \geq 1 - \varepsilon.$$

Let the mapping $size : \{0, 1\}^n \to \mathbb{N}$ associate a natural number $size(f)$ with a target concept $f \in F$ that is a measure of complexity of f under a target representation. If a concept class F has infinite cardinality, i.e., an infinite concept class, the Vapnik–Chervonenkis dimension $VC_{dim}(F)$ is a general measure of its complexity introduced by Vapnik and Chervonenkis in their seminal studies [117, 118]. This measure is defined as follows.

Definition 2.6.1 ([64]) The Vapnik–Chervonenkis dimension of a concept class F denoted by $VC_{dim}(F)$ is the largest cardinality of a set of examples S meeting the following requirement. For every subset $U \subseteq S$, a concept $f \in F$ exists so that $U = S \cap f$.

The importance of this definition lies primarily in finding a lower bound/ an upper bound on the number of examples required by a learning algorithm. Before elaborating on this lower bound, we should shift our focus to the definition of a learner.

A learner is a polynomial-time algorithm denoted by A, which is given labeled examples $(c, f(c))$, where $c \in C_n$ and $f \in F_n$. The examples are drawn independently according to the distribution D. Now we can define strong and weak PAC learning algorithms.

Definition 2.6.2 An algorithm A is called a **strong** PAC learning algorithm for the target concept class F, if for any $n \geq 1$, any distribution D, any $0 < \varepsilon, \delta < 1$, and any $f \in F_n$ the following holds. When A is given a polynomial number of labeled examples, it runs in time polynomial in n, $1/\varepsilon$, $size(f)$, $1/\delta$, and returns an ε-approximator for f under D, with probability at least $1 - \delta$.

The weak learning framework was developed to answer the question whether a PAC learning algorithm with constant but insufficiently low levels of ε and δ can be useful at all. This notion is defined as follows.

Definition 2.6.3 For some constant $\delta > 0$ let algorithm A return with probability at least $1 - \delta$ an $(1/2 - \gamma)$-approximator for f, where $\gamma > 0$. A is called a **weak** PAC learning algorithm, if $\gamma = \Omega(1/p(n, size(f)))$ for some polynomial $p(\cdot)$.

Algorithm 1 The skeleton of a Boosting algorithms.

Require: Weak PAC learner WL, $0 < \varepsilon, \delta < 1, 0 < \gamma \leq 1/2$, polynomial number of examples, i
    that is the number of iterations
Ensure: Hypothesis h that is an ε-approximator for f

1: $D_0 = D$, use WL to generate an approximator h_0 for f under D_0
2: $k = 1$
3: **while** $k \leq i - 1$ **do**
4:     Build a distribution D_k consisting of examples, where the previous approximators
    $h_0, \cdots, h_{k-1}$ can predict the value of f poorly
5:     use WL to generate an approximator h_k for f under D_k
6:     $k = k + 1$
7: **od**
8: Combine the hypotheses $h_0, \cdots, h_{i-1}$ to obtain h, where each h_i is an $(1/2 - \gamma)$-approximator
    for f under D_i, and finally h is an ε-approximator for f under D
9: **return** h

A PAC learner A, in general, requires at least the following number of examples [15]:

$$\Omega \left(\frac{1}{\varepsilon} \log \frac{1}{\delta} + VC_{dim}(F) \right).$$

Note that this lower bound on the number of examples has been proved for the condition $0 < \varepsilon < 1/2$ met by PAC learning algorithms in general. We stress that the above mentioned lower bound has been improved for strong PAC learning algorithms in [25]. Nonetheless, it is not applicable in the case of weak PAC learners.

Last but not least, it should be noted that the equivalence of weak PAC learning and strong PAC learning has been proved by Freund and Schapire in the early nineties in their seminal papers [27, 99]. For that purpose *boosting* algorithms have been introduced.

Definition 2.6.4 An algorithm B is called a boosting algorithm if the following holds. Given any $f \in F_n$, any distribution D, $0 < \varepsilon, \delta < 1$, $0 < \gamma \leq 1/2$, a polynomial number of labeled examples, and a weak learning algorithm WL returning an $(1/2 - \gamma)$-approximator for f, then B runs in time, which is polynomial in n, $size(f), 1/\gamma, 1/\varepsilon, 1/\delta$ and generates with probability at least $1 - \delta$ an ε-approximator for f under D.

The construction of virtually all existing boosting algorithms is based primarily on the fact that if WL is given examples drawn from any distribution D', WL returns a $(1/2 - \gamma)$-approximator for f under D'. At a high-level, the skeleton of all such boosting algorithms is shown in Algorithm 1.

PAC Learning of LTFs with Perceptron Algorithm

Several studies have focused on the PAC learning of an unknown LTF from labeled examples by applying the Perceptron algorithm (for an exhaustive survey see [104]). This issue plays an important role in our framework. Here we briefly describe how the Perceptron algorithm, as an online algorithm, can be converted to a PAC learning algorithm, following the conversion procedure defined in [104].

The learner has access to an Oracle EX, providing labelled examples. By calling EX successively, a sequence of labeled examples is obtained and fed into the online algorithm. Hypotheses generated by the algorithm are further stored. At the second stage, the algorithm again calls EX to receive a new sequence of labeled examples. This new sequence is used to calculate the error rate of the hypotheses stored beforehand. The output of the procedure is a hypothesis with the lowest error rate. Let ε and δ be the accuracy and the confidence levels of the obtained PAC learning algorithm. Suppose that N_{mis} is the upper bound of the mistakes made by the original online algorithm for the concept class F. The following theorem is proved by Littlestone [66].

Theorem 1: *Suppose that the online algorithm* $\mathbb{A}_{on}$ *improves its hypothesis, only when its prediction and the received label of the example do not agree. The total number of calls that the obtained PAC algorithm* $\mathbb{A}$ *makes to* EX *is* $O\left(1/\varepsilon(\log 1/\delta + N_{mis})\right)$.

From the convergence theorem of the Perceptron algorithm and Theorem 1, it is straightforward to prove the following corollary [104]:

Corollary 2.6.1 *Let the concept class* F_n *over the instance space* $C_n = \{0, 1\}^n$ *be the class of LTFs such that the weights* $\omega_i \in \mathbb{Z}$, *and* $\sum_{i=1}^{n} |\omega_i| = p(n)$. *Then the Perceptron algorithm can be converted to a PAC learning algorithm running in time* $p(n, 1/\varepsilon, 1/\delta)$.

Chapter 3
PAC Learning of Arbiter PUFs

> *[In our ML attack against PUFs, e.g., Arbiter PUFs] The*
> *iteration is continued until we reach a point of convergence [...].*
> *If the reached performance after convergence on the training set*
> *is not sufficient, the process is started anew.*
>
> [95]

This chapter is based on [36], slightly modified to fit within the structure of this thesis. The current chapter aims at establishing a new representation of Arbiter PUFs that reflects the physical properties of these PUFs. This representation enables us to come up with new results on the learnability of Arbiter PUFs for given levels of accuracy and final model delivery confidence. This is in contrast to previous studies (e.g., [95]), where it is not clear whether after the learning phase, a model of the Arbiter PUF with the desired level of accuracy would be delivered by the machine learner. Finally, we discuss the importance of our framework from the practical point of view.

Overview of this chapter: After an introduction presented in Sects. 3.1 and 3.2 a mathematical representation of Arbiter PUFs is introduced. Section 3.3 explains how the challenge-response behavior of an Arbiter PUF can be learned for given levels of accuracy and confidence. Section 3.4 presents a brief overview of the relevant literature. In Sect. 3.5, additional important aspects of our methodology are discussed.

3.1 Introduction

The core idea behind the design of Arbiter PUFs is to exploit the delay differences between symmetrically designed electrical paths on a silicon chip to generate a somehow random but unique response [61, 62]. Similar to all families of PUFs,

© Springer International Publishing AG 2018
F. Ganji, *On the Learnability of Physically Unclonable Functions*, T-Labs
Series in Telecommunication Services, https://doi.org/10.1007/978-3-319-76717-8_3

unclonability and *unpredictability* are the main requirements of the Arbiter PUF family [8, 88]. However, contrary to these basic requirements, previous work in the literature introduced different successful attacks on Arbiter PUFs. Interestingly enough, the fragility of their security has been discussed in the first papers introducing the Arbiter PUFs [39, 62]. Nevertheless, it was assumed that launching attacks against Arbiter PUFs could not be feasible. These attacks against Arbiter PUFs can be classified into two categories: side channel attacks and modeling (i.e., ML) attacks. The former type of attacks exploits the side channel information, such as photonic emissions and electromagnetic radiations, to physically characterize an Arbiter PUF [23, 24, 109]. On the other hand, modeling attacks require only a subset of CRPs to build a mathematical model of the Arbiter PUF, which later can predict the response of that Arbiter PUF, with some probability [61, 95]. As being non-invasive, modeling attacks can be more cost and time effective in comparison to side channel attacks.

Although an increase in the number of PUF stages can reduce the effectiveness of modeling attacks [63, 107], by utilizing more advanced machine-learning tools an attacker can still break the security of Arbiter PUFs [96]. On the one hand, it has been verified experimentally that the number of CRPs required for a successful attack increases exponentially with the number of stages [68]. On the other hand, it has been empirically shown that the relation between this number of CRPs and the number of stages can be described by a linear function [95]. This raises the natural question to what degree of precision an attacker can model an Arbiter PUF, with an arbitrary number of stages. In other words, how many CRPs are required to model the PUF for given levels of *accuracy* and final model delivery *confidence*. Unfortunately, this issue has not been completely addressed in the literature so far, and thus modeling attacks rely only on trial and error or heuristic approaches.

This chapter presents a well-defined mathematical representation of Arbiter PUFs. In the literature, with regard to the linear additive model underlying the construction of Arbiter PUFs, a common, widely applied representation of these PUFs has been established cf. [39, 62, 95]. Nevertheless, we believe that the new representation introduced in our work helps to not only deepen our understanding of security weaknesses of Arbiter PUFs, but also draw an analogy between Arbiter PUFs and one of the prominent issues in ML theory, namely, learning a regular language. Seen from this new perspective, interesting and important aspects of the design of the Arbiter PUF family are recognized in our work. Moreover, based on this representation, we introduce a polynomial-time learning algorithm that provably learns the challenge-response behavior of an arbitrary Arbiter PUF, for given, prescribed levels of accuracy and confidence. We show how the levels of accuracy and confidence of a model are related to the number of collected CRPs and the number of stages of an Arbiter PUF as well as the maximum variation of delay values. We prove that the maximum number of CRPs required for our attack is polynomial in the number of stages. Finally, we evaluate the time complexity of our learning algorithm and prove that it is polynomial in the number of stages, the maximum variation of delays inside the stages, and levels of accuracy and confidence. The main contributions of this chapter are as follows:

A learning algorithm for prescribed levels of accuracy and confidence. Based on a new mathematical representation for Arbiter PUFs, we introduce a learning algorithm, which is able to model the PUF for given levels of accuracy and confidence.

Discretization of real-valued delay values. The key idea behind the new representation of Arbiter PUFs is to define a set of proper *integer* delay values, which contains the statistically relevant delay values of an Arbiter PUF. We first explain that these delay values are distributed within a limited interval. Secondly, demonstrate how we can discretize the different delay values with regard to the limited precision of the Arbiter placed at the end of its respective chain. Finally, due to these facts, we present a mapping between these discrete delay values to a fitting set of integer values.

Calculation of the maximum number of CRPs required for our new attack. In order to learn the challenge-response behavior of an Arbiter PUF for given levels of accuracy and confidence, we prove that the maximum number of CRPs required for launching our attack is polynomial in the number of stages composing an Arbiter PUF. Besides that, the impact of the limited variation of the delays on the learnability of an Arbiter PUF is discussed.

Evaluation of the time complexity of the new attack. Finally, we evaluate the time complexity of our learning algorithm. Our proofs reveal that the running time of the proposed learning algorithm is polynomial in the length of the given Arbiter PUF (i.e., the number of stages), the maximum variation of delays and the given levels of accuracy and confidence.

3.2 Representing Arbiter PUFs by DFAs

It is known that in order to PAC-learn a target concept it is necessary to come up with a polynomial-sized representation of that [56]. Therefore, to PAC learn the intrinsic challenge-response behavior of a given PUF, a polynomial-size representation of its behavior is required. Here, we aim to derive such a concise representation that can afterwards be used to provide a PAC learning algorithm, which runs in polynomial time and learns the unknown challenge-response behavior of an Arbiter PUF for predefined levels of accuracy and confidence.

For Arbiter PUFs, similar to other types of PUFs, we stick to the definition of PUFs as input-output mappings introduced in Chap. 2 (see Sect. 2.2). An Arbiter PUF consists of multiple switch blocks, so-called *stages*, connected in a chain terminated by an arbiter, see Fig. 3.1. A challenge is a string $c = c[1] \cdots c[n]$ of n bits, where each bit (e.g., $c[i]$) is fed into a single stage (e.g., ith stage). The signal propagates through the direct paths inside the ith stage if $c[i] = 0$, otherwise the crossed paths are utilized. Let $\mathcal{B}_i$ denote a random variable related to the delay within the ith stage. The realizations of the variables $\mathcal{B}_i$ in an Arbiter PUF are certain $\overline{\beta}_{i,1}, \overline{\beta}_{i,2}, \overline{\beta}_{i,3}$, and $\overline{\beta}_{i,4}$. Here $\overline{\beta}_{i,1}$ and $\overline{\beta}_{i,2}$ are the delays of the upper and lower direct paths in the ith

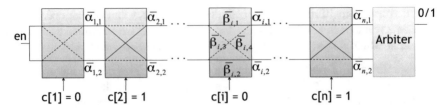

Fig. 3.1 Schematic of an Arbiter-PUF

stage, respectively, see Fig. 3.1. On the other hand, the delays of the upper and lower crossed paths in the ith stage are $\overline{\beta}_{i,3}$ and $\overline{\beta}_{i,4}$, respectively. $\mathcal{B}_i$ follows a Gaussian distribution with the mean μ_i and the deviation σ_i, cf. [62].

We define $\mathcal{A}_i$ as a random variable corresponding to the total delay between the enable point and the outputs of the ith stage of the PUF. Following the linear additive model of the Arbiter PUF, cf. [61], we have

$$\mathcal{A}_i = \sum_{k=1}^{i} \mathcal{B}_k.$$

The realizations of the partial sums $\mathcal{A}_i$ at the outputs of the ith stage are denoted by $\overline{\alpha}_{i,j}$, where j represents the upper and lower output (i.e., $j = 1$ for upper and $j = 2$ for lower output), see Fig. 3.1. The arbiter at the end of the PUF chain has a precision $\gamma > 0$, and compares the arrival times of signals on the upper and lower paths (i.e., $\overline{\alpha}_{n,1}$ and $\overline{\alpha}_{n,2}$). More formally, we assume that the output of the Arbiter is "1", if the delay difference at the outputs of the last stage is

$$\Delta = \overline{\alpha}_{n,1} - \overline{\alpha}_{n,2} > \gamma,$$

whereas it is "0" if $\Delta < -\gamma$. The metastable condition, where $|\overline{\alpha}_{n,1} - \overline{\alpha}_{n,2}| < \gamma$, will be discussed later in Sect. 3.5. Finally, in order to take into account the impact of different path configurations for $\overline{\alpha}_{n,1}$ and $\overline{\alpha}_{n,2}$, we define the single bit

$$u_i = \bigoplus_{k=1}^{i} c[k],$$

related to the history of the paths that the signal follows. This single bit u_i has often been called *parity of challenge bits* (cf. [62, 75]).

To derive a polynomial-sized representation of Arbiter PUFs, we first show how the real delay values of an Arbiter PUF can be *mapped* to a finite set of integer values.

3.2.1 Discretization Process of Delay Values

As mentioned before, the delay differences in the stages of an Arbiter PUF are caused by variations in the manufacturing processes. It is known that $\mathcal{B}_i$ (the aforementioned random variable describing the delay of the ith stage) follows a Gaussian distribution, cf. [62]. Thus, $\mathcal{B}_i \sim N(\mu_i, \sigma_i)$, and its Probability Density Function (PDF) $f_{\mathcal{B}_i}(\overline{\beta}_i)$ is given by

$$f_{\mathcal{B}_i}(\overline{\beta}_i) = \frac{1}{\sigma_i \sqrt{2\pi}} e^{-(\overline{\beta}_i - \mu_i)^2 / 2\sigma_i^2}.$$

The mean μ_i is often reported by manufactures as the nominal propagation delay of the utilized multiplexer, and the standard deviation σ_i is caused by the variations in the manufacturing process, cf. [75]. As 99.7% of Gaussian distributed values lie within the range of three standard deviation away from the mean value, the realizations $\overline{\beta}_{i,1}$, $\overline{\beta}_{i,2}$, $\overline{\beta}_{i,3}$, and $\overline{\beta}_{i,4}$ are drawn from an interval, whose length is 6σ, see Fig. 3.2.

With regard to the additive linear model of the Arbiter PUF we have $\mathcal{A}_i = \sum_{k=1}^{i} \mathcal{B}_i$, where $\mathcal{A}_i$ is the random variable, which shows the total propagation delays at the outputs of the ith stage. Therefore, the PDF of the total propagation delays at the outputs of each stage are the convolution of all PDFs of the previous stages [85], i.e.,

$$f_{\mathcal{A}_i}(\overline{\alpha}_i) = f_{\mathcal{B}_i}(\overline{\beta}_i) * f_{\mathcal{B}_{i-1}}(\overline{\beta}_{i-1}) * \cdots * f_{\mathcal{B}_1}(\overline{\beta}_2) * f_{\mathcal{B}_1}(\overline{\beta}_1).$$

As all delays in each stage follow the normal distribution, $\mathcal{A}_i$ also follows the normal distribution. Hence, in an Arbiter PUF of length n, if we assume that $\mu_1 = \mu_2 = \cdots = \mu_n = \mu$ and $\sigma_1 = \sigma_2 = \cdots = \sigma_n = \sigma$, we can write

$$f_{\mathcal{A}_i}(\overline{\alpha}_i) = \frac{1}{\sigma \sqrt{2i\pi}} e^{-(\overline{\alpha}_i - i\mu)^2 / 2i\sigma^2}.$$

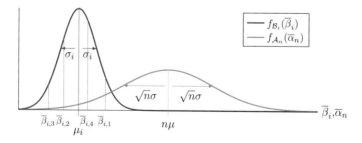

Fig. 3.2 The distribution of $\mathcal{B}_i$ (blue) with the mean μ_i and deviation σ_i. Four examples of possible realization of $\mathcal{B}_i$ are $\overline{\beta}_{i,1}, \overline{\beta}_{i,2}, \overline{\beta}_{i,3}$, and $\overline{\beta}_{i,4}$, which correspond to four delays at the ith stage. The distribution of total propagation delay from the enable point to the outputs of last stage in an Arbiter PUF with length n (red), mean $n\mu$, and deviation $\sqrt{n}\sigma$

As a result, the random variable $\mathcal{A}_n$, corresponding to the total propagation delays at the last stage, will have the mean $n\mu$ and the standard deviation $\sqrt{n}\sigma$, see Fig. 3.2. Therefore, we can assume that all statistically relevant delay values lie within a limited interval, whose length is $6\sqrt{n}\sigma$ (i.e., within three standard deviation away from the mean value $n\mu$).

It is obvious that the delay differences are real numbers. However, the arbiter at the end of the chain provides only a limited precision in terms of comparing the total propagation delays of two paths [73]. Hence, it can compare two signals with delay differences only above a certain threshold, say $\gamma > 0$. As a result, the actual number of different delay values, which can be observed and compared by the arbiter are limited. Due to this fact, all propagation delays at the output of each stage (i.e., $\overline{\alpha}_{i,j}$, where $0 \leq i \leq n$ and $1 \leq j \leq 2$) can be mapped to integer values. A mapping $f : \mathbb{R} \mapsto \mathbb{Z}$ is defined as follows. For all $\overline{\alpha} \in [n\mu - 3\sigma\sqrt{n}, \; n\mu + 3\sigma\sqrt{n}]$, we have

$$f(\overline{\alpha}) = \left\lceil \frac{\overline{\alpha} - n\mu + 3\sigma\sqrt{n}}{\gamma} \right\rceil.$$

It is straightforward to show that all real delay values lying in the interval $[n\mu - 3\sigma\sqrt{n}, \; n\mu + 3\sigma\sqrt{n}]$ are mapped to integer values between 0 and M, where

$$M = \left\lceil \frac{6\sqrt{n}\sigma}{\gamma} \right\rceil.$$

Note that 0 and M correspond to the minimum and the maximum of the statistically relevant real values, respectively. The mapped values are denoted by $\alpha_{i,j}$, i.e., we have $\alpha_{i,j} \in \mathbb{Z}$. In this case, the response of the arbiter is "1" if $\alpha_{n,1} - \alpha_{n,2} \geq 1$, whereas it is "0" if $\alpha_{n,2} - \alpha_{n,1} \geq 1$. The arbiter is in the metastable condition, if $|\alpha_{n,1} - \alpha_{n,2}| = 0$. Moreover, clearly, $\overline{\beta}_{i,j}$, where $0 \leq i \leq n$ and $1 \leq j \leq 2$, can be mapped to an integer value $\beta_{i,j}$ lying in the interval $[0, M]$ (see Sect. 4.2).

On the basis of this mapping, we introduce a DFA-based representation of an Arbiter PUF.

3.2.2 Building a DFA Representing an Arbiter PUF

Consider a PUF, whose challenge-response functionality is given by the mapping f_{PUF} defined in Sect. 2.2. Let us define $L_{f_{\mathrm{PUF}}} := \{c \in \mathcal{C} \mid f_{\mathrm{PUF}}(c) = 1\}$. We have $L_{f_{\mathrm{PUF}}} \subseteq \{0, 1\}^n \subseteq \Sigma^*$, where $\Sigma = \{0, 1\}$. Hence, $L_{f_{\mathrm{PUF}}}$ can be thought as being the accepted language of a certain automaton. It accepts those strings $c \in \mathcal{C}$, whose length is n and $f_{PUF}(c) = 1$.

In order to build an automaton A, we will use the notation of the *integer* PUF propagation delays $\alpha_{i,j}$, see Sect. 3.2.1. Figure 3.3 illustrates the central idea of our automaton construction. After reading the first challenge bit applied to the first stage, A transits from q_0 (the initial state) to either $q_{1,1}$ or $q_{1,2}$, depending on whether

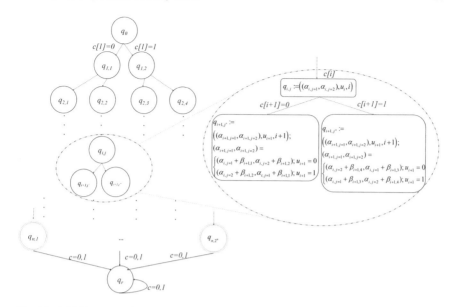

Fig. 3.3 A DFA representing an Arbiter PUF

$c[1] = 0$ or $c[1] = 1$. Here $q_{1,1}$ and $q_{1,2}$ correspond to tuples $((\alpha_{1,1}, \alpha_{1,2}), u_1, i = 1)$, where the first elements of them are ordered pairs of possible delays at the output of the first stage. Note that due to the physical characteristics of the Arbiter PUF, i.e., the signal propagated through either the direct or crossed path, two different pairs of delays are obtained, and included in $q_{1,1}$ and $q_{1,2}$. The second element of these tuples is u_1 which *memorizes* the first ith bits of the challenge: for the first stage we have $i = 1$ and thus, $u_1 = c[1]$. Finally, the third element of these tuples represents the depth in which the respective nodes are, e.g., $i = 1$ for $q_{1,1}$ and $q_{1,2}$. This component *counts* the number of bits of the input c, which have been consumed so far.

In order to further elaborate on the definition of $q_{i,j}$, we provide an example on how the DFA transits from $q_{i,j}$ to $q_{i+1,j'}$ when reading $c[i + 1] = 0$, see Fig. 3.3. As can be seen from the figure, when $u_{i+1} = 0$, the first entry of the pair $(\alpha_{i+1,1}, \alpha_{i+1,2})$ contains the delays on the upper paths (i.e., $\alpha_{i+1,1} = \alpha_{i,1} + \beta_{i+1,1}$), whereas for $u_{i+1} = 1$ it is composed of the delays on the lower paths, i.e., $\alpha_{i+1,1} = \alpha_{i,2} + \beta_{i+1,2}$. Hence, using u_{i+1}, the correct sum is defined. What has been explained here is applicable for the other stages of the Arbiter PUF as well, and consequently, the other states of A can be defined similarly.

The further and crucial characteristics of this DFA representing an Arbiter PUF are as follows. In Fig. 3.3, the states $q_{n,1}, \ldots, q_{n,2^n}$ define the *possible* accepting states in the following way. As mentioned in Sect. 3.2.1, assume that $f_{\text{PUF}}(c) = 1$, if $\alpha_{n,1} - \alpha_{n,2} > 1$, and when $\alpha_{n,2} - \alpha_{n,1} > 1$, $f_{\text{PUF}}(c) = 0$. For a challenge string $c = c_1 c_2 \cdots c_n$, A accepts c if there is a sequence of states such that $r_0 = q_0$, $r_i = \delta(r_{i-1}, c_i)$ with $1 \leq i \leq n$ and $r_n \in F$, where

$$F = \{q_{n,k} \mid q_{n,k} = ((\alpha_{n,1}, \alpha_{n,2}), u_n, n) \text{ s.t. } \alpha_{n,1} - \alpha_{n,2} \geq 1, 1 \leq k \leq 2^n\}.$$

All other states, not being defined as accepting states, are of course rejecting. Although it is possible that $\alpha_{n,1} = \alpha_{n,2}$, in reality it may not occur (see Sect. 3.5 for more details), thus we can safely exclude this case. The special rejection state q_r is reached after reading further bits following the nth bit of a given input. After reaching the rejecting state, reading any further bit results in staying in q_r. We should stress that since challenges are n-bit strings, all longer strings are rejected, as well as shorter strings.

As it is evident from the above discussion, the size of A constructed based upon the real-valued total delays at the output of each stage of the Arbiter PUF is exponential in n. Consequently, if we represent an Arbiter PUF by this A, the output of a learning algorithm is a hypothesis h with $h \in H_n$ that could not be evaluated in polynomial time. However, having a closer look, we prove that this DFA has indeed only a size which is polynomial in n, and can be used to PAC-learn an Arbiter PUF. To shrink A, we use the results of the discretization process of the real delay values.

As mentioned in Sect. 3.2.1, the total delay values can be mapped to the integer values lying in the finite set $[0, M]$, where M can be regarded as a constant, independent of n. Therefore, the number of possible values of $\alpha_{i,j}$ for $1 \leq i \leq n$ and $1 \leq j \leq 2$ is $M + 1$. For a given depth i, the number of possible pairs of delays on the upper and lower paths (i.e., $(\alpha_{i,1}, \alpha_{i,2})$) is thus at most $(M + 1)^2$. Consequently, the number of distinguishable pairs, i.e., corresponding to the different ordered pairs

Fig. 3.4 The shrunk DFA representing an Arbiter PUF. Note that the size of this DFA is clearly polynomial in n

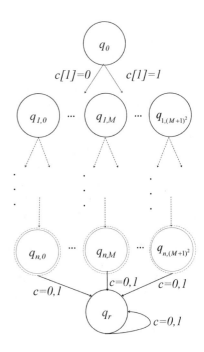

of sums in each level of A cannot exceed $(M + 1)^2$, and thus the total number of distinguishable states is limited by

$$O(n(M + 1)^2).$$

Collapsing the indistinguishable states, a much smaller DFA as shown in Fig. 3.4 is obtained.

3.3 PAC Learning of Arbiter PUFs

With the help of the polynomial-size DFA built in Sect. 3.2, we can now describe a PAC learning algorithm that efficiently learns the challenge-response behavior of a given Arbiter PUF. Such a PAC learning algorithm can be derived by adopting and modifying the algorithms presented in the literature [4, 56]. The algorithm presented later by us can be seen as an adapted version of what has been proposed by Angluin [4]. Here we only describe the algorithm briefly, and refer the reader to [4] for further details.

A given PUF provides the learner with access to the Oracle $EX := f_{PUF}$ (see Fig. 3.5). Angluin's algorithm can be applied even in the case that the oracle EX outputs examples with different lengths. However, in our case, our Oracle EX provides labeled examples, whose length are exactly n. This means that the PAC learning problem applied by us can be thought of as being a simplified version of the more general problem solved in [4]. Furthermore, note that our algorithm is designed to learn those challenges yielding a "1" at the output of the final arbiter of the given PUF, positives examples given by EX.

As illustrated in Fig. 3.5, the length of the examples n, the maximum delay value M, and the levels for the accuracy and confidence are provided as further inputs to the algorithm. The main steps of the PAC learning algorithm are depicted in Algorithm 1. At the first stage, h_0 contains λ and no string c is accepted. Since all examples with $f_{PUF}(c) = 0$ are rejected, and h_0 will be modified only after receiving a positive example, without loss of generality, we assume that the first example is positive. For the ith example, the algorithm examines whether h_{i-1} is consistent. Then h_{i-1} is updated, if it is not consistent with the example. The procedures of checking the consistency and updating a hypothesis are described extensively in [4, 56], and are not further discussed by us. Moreover, the proof of the correctness of the above algorithm, A, is also presented in [4], and the reader is referred to that for more

Fig. 3.5 Block diagram of the PAC learning algorithm

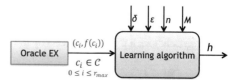

details. It is ensured that A makes at most r_{max} calls to the oracle EX, and the final output h is an $\varepsilon/2$-approximation of S^1 (i.e., set of positive examples), as shown by the following theorem.

Theorem 3.3.1 *Let $N := O(nM^2)$ that represents the number of live states, then A returns a hypothesis h after at most $O(N + (1/\varepsilon)(N \log(1/\delta) + N^2))$ calls to EX, and with probability at least $(1 - \delta/2)$, h is an $\varepsilon/2$-approximation of S^1.*

As all the details of the proof are already given in [4], we elaborate only on a few interesting points from that. The proof is based on the maximum number of calls that A makes to EX, r_{max}, and can be calculated as follows cf. [4]. According to the PAC model (see Sect. 2.6), with probability at least $1 - \delta/2$, A should return a $1 - \varepsilon/2$ accurate hypothesis. Hence, it is straightforward to show that if A has tested i conjectures so far, the number of calls to the oracle EX is

$$\lceil r_i \rceil = \frac{1}{\varepsilon}\left(\ln(1/\delta) + \log_2(i + 1)\right).$$

As at most $(N - 1)$ different h_k have to be tested, we have

$$r_{max} = \sum_{i=0}^{(N-2)} (r_i + 1)$$

$$= (N - 1) + 2/\varepsilon\left((N - 1)\ln(1/\delta) + \ln(2)\sum_{i=0}^{(N-2)} (i + 1)\right),$$

which results in

$$O\left(\left(1 + \frac{2}{\varepsilon}\ln(1/\delta)\right)n(M + 1)^2 + \frac{2}{\varepsilon}n^2(M + 1)^4\right).$$

Now we calculate the probability that the output h is not a $\varepsilon/2$-approximation of S^1. As in [4] we have

$$\Pr[error(h) > \varepsilon/2] = \sum_{i=0}^{(N-2)} (1 - \varepsilon/2)^{r_i},$$

which yields that

$$\Pr[error(h) \leq \varepsilon/2] \geq 1 - \frac{\delta}{2}.$$

The time consumed by A is bounded by the time spent on searching for strings, which distinguish two states in h_k. Since the maximum size of a set containing all possible strings is $|\Sigma| \cdot |A_{\text{PUF}}| + 1$, which is $2(n \cdot (M + 1)^2) + 1$, and the related steps (step 4

Algorithm 2 PAC learning algorithm A for f_{PUF}.

Require: A list of $(c_i, f_{\text{PUF}}(c_i))$, (ε, δ), n, and M
Ensure: h

1: $h_0 := \lambda$
2: Wlog. let the first positive example be $(c_1, 1)$
3: $k = 0, i = 1$
4: Examine the consistency of h_k
5:   **if** h_k is not consistent with c_i **then**
6:     Update h_k
7:     $k = k + 1$
8:   **fi**
9: $i = i + 1$
10: Let $r_i := \lceil 2/\varepsilon(\ln(1/\delta) + (i+1)\ln 2) \rceil$
11: **while** $r_i \leq r_{max}$ **do**
12:   Proceed with the next example
13:   Do steps 4-8
14: **od**
15: $h := h_k$
16: **return** h

---

to 8 in Algorithm 1) should be repeated at most r_{max} times, the time complexity of A is bounded by

$$O\left((1 + (2/\varepsilon)\ln(1/\delta))n^2(M+1)^4 + (2/\varepsilon)n^3(M+1)^6\right).$$

Thus, we conclude that S^1 can be efficiently learned (for given accuracy and confidence levels), and the respective algorithm runs indeed in time polynomial in n, M, $1/\varepsilon$ and $1/\delta$.

Note on the Definition of the Accuracy Level ε

Last but not least, we should further elaborate on the definition of the accuracy level ε, and more specifically, on the impact of building an erroneous DFA. In Sect. 2.6, ε is defined as the error of hypothesis delivered by the PAC learning algorithm. This error can occur, when either deriving the representation of an Arbiter PUF or learning from membership queries. The latter case is well studied in the relevant literature cf. [4]. Here we focus on the error that may occur while building the DFA-based representation of an Arbiter PUF. As mentioned, PAC learning of an Arbiter PUF under DFA representation is the direct consequence of delay discretization. Therefore, the error of discretization process should be taken into account.

When calculating M (the maximum delay variation), we assume that all the statistically relevant delay values are lying within a limited interval $[n\mu - 3\sigma\sqrt{n}, n\mu + 3\sigma\sqrt{n}]$, corresponding to 99.7% of the total values. It can be thought that the values lying outside of this interval are not considered, which may result in an error. Hence the definition of ε should be refined to reflect the impact of this error. To this end, we claim that the error of discretization process can be limited to $\varepsilon/4$, that is equal to the error of learning. Therefore, even in the case of an erroneous DFA representation, the error of the delivered hypothesis does not exceed $\varepsilon/2$.

We begin with the calculation of the probability that the delay values are lying outside the interval $[n\mu - 3\sigma\sqrt{n}, \ n\mu + 3\sigma\sqrt{n}]$. That is

$$\Pr\left(|\overline{\alpha}_n - n\mu| \leq 3\sigma\sqrt{n}\right) = \text{erf}(\frac{3}{\sqrt{2}}) = 0.9973.$$

Now we define $1 - \Pr\left(|\overline{\alpha}_n - n\mu| \leq 3\sigma\sqrt{n}\right) \leq \varepsilon/4$ that yields $0.01 \leq \varepsilon$. Comparing this result with the condition defined in PAC model ($0 < \varepsilon < 1$), it can be understood that in the case of erroneous DFA, the error of the delivered hypothesis is lower bounded by $\varepsilon/2 = 0.5\%$.

3.4   Comparison with Related Work

We have proved that the Arbiter PUF family is subject to PAC learning attacks and presented an algorithm, which predicts the output of an Arbiter PUF for given ε and δ. The notion of PAC learning has already been used in the context of PUFs by Hammouri et al. [45] to assess the security of their proposed scheme. In order to do so, they represent a noiseless Arbiter PUF as an LTF [7], as first proposed in [39, 62]. This implicitly leads to the conclusion that a noiseless Arbiter PUF is PAC-learnable, as the Vapnik–Chervonenkis dimension of the proposed representation (LTF) for a noiseless Arbiter PUF is equal to $n + 1$ (for more details and a proof of this, cf. [7]). Consequently, a noiseless Arbiter PUF is in principle PAC-learnable under LTF-based representations. However, this has not been revealed in [45], since the discretization of real-valued delays was neither explored nor known at that time. And indeed only this process enables the Perceptron algorithm to PAC-learn LTFs representing Arbiter PUFs, cf. [7]. To the best of our knowledge, our framework is the first algorithm that provably learns an arbitrarily chosen Arbiter PUF under a well-established representation, for prescribed levels of accuracy and confidence, in polynomial time.

Since the mathematical proof is proposed here, it seems redundant to conduct further experiments or simulations that provide a proof of concept. Nevertheless, for the sake of completeness, we compare our theoretical findings with experimental results reported in [95]. Note that the hypothesis class of Logistic Regression algorithm (LR) applied in [95] can be discretized to obtain a finite hypothesis class [105]. Moreover, thanks to our discretization process, the loss function of LR is also bounded, hence, it can be converted to a PAC learning algorithm. In this case, the maximum number of CRPs required to launch an attack is polynomial in n, $1/\varepsilon$ and $1/\delta$. According to above discussion, we can attempt to compare the number of CRPs required by our algorithm and LR. We have proved that the number of CRPs required by our PAC learning framework is polynomial in n, M, $1/\varepsilon$ and $1/\delta$. This has been only partially and empirically verified in previous work, e.g., [95], where the impact of M and δ

has not been known. They have shown that when n is increased from 64 to 128, the number of CRPs required to model the PUF for $\varepsilon = 0.01$ increases almost linearly. However, as mentioned before, it is not ensured that the final model may be delivered after the learning phase. In other words, it is not guaranteed that for given levels of accuracy and confidence, an Arbiter PUF can be modeled by collecting such linearly increasing number of CRPs. This is contrary to our algorithm, where the number of CRPs is polynomial in n and the final model is delivered with probability at least $1 - \delta$. Furthermore, and more crucially, by relying primarily on the experimental results the authors of [95] have *estimated* the number of CRPs and the time required for launching the attack in general. In contrast to this, we have *calculated* the number of CRPS and the time complexity of our algorithm based on our mathematical proof.

3.5 Practical Considerations

This section devoted to the lessons learned from the practice that enable us to prove the vulnerability of Arbiter PUFs to PAC learning attacks.

3.5.1 The Important Role of M

The essential aspect of our framework is related to our DFA-based representation established with regard to the observation that the delay values can be mapped to a finite set of integer values. Here we further elaborate on this observation. It has been demonstrated that for a Xilinx Virtex-5 FPGA [73] the maximum delay deviation of each inverter used in the PUF chain is 9 ps for both cases, i.e., direct and crossed paths on average. This delay difference is virtually in line with what has been observed by authors of [72], where for 12 XC5VLX110 chips (Xilinx Virtex-5 family) the delay deviation is smaller than 10 ps on all chips. Assuming $6\sigma = 10$ ps, the maximum variation of delay at the end of the PUF chain consisting of n stages is $10\sqrt{n}$ ps. However, this value has to be divided by the precision of the arbiter to calculate the maximum value M, cf. Sect. 3.2.1. For an absolute precise arbiter, the precision γ can be thought as being infinitesimal, i.e., $\gamma \rightarrow 1/\infty$, see [73]. Nevertheless, the precision of the arbiter is reported to be only in the range of 2.5 ps for a Xilinx Virtex-5 FPGA, cf. [73]. As an example, under the assumption that $n = 128$,

$$M = \lceil 6\sigma\sqrt{n}/\gamma \rceil = \lceil 10 \cdot \sqrt{128}/2.5 \rceil = 46.$$

Therefore, the size of the collapsed DFA would be only 282,752 states. This is far less than the size of $O(2^{128})$ states, which would be obtained without considering the limited variation of the delays.

Naturally, to make PAC learning of a concrete PUF less effective, it is tempting to construct a PUF with a very large M. Theoretically, the maximum delay value M can be increased by enlarging the manufacturing deviations, and also using more precise arbiters. However, the deviation σ cannot be arbitrarily large on real production chips. For instance, when increasing σ, a Field Programmable Gate Array (FPGA) cannot be utilized anymore. Moreover, with regard to the higher cost, the PUF manufacturers cannot arbitrarily increase the precision of the arbiter. Due to these limitations, arbitrarily large M's can be excluded in practice for FPGAs and standard CMOS process devices.

3.5.2   Dealing with the Metastable Condition

In Sect. 3.2.2, we have stated that in practice it may rarely happen that $\alpha_{n,1} = \alpha_{n,2}$. This can be explained by the fact that this equation represents the possible metastable condition of the Arbiter PUF, when the output of the arbiter is not persistent for a certain challenge c. Note that this metastablity of the Arbiter PUF must have been already solved by the PUF manufacturer. Moreover, aiming at PAC learning of an Arbiter PUF under a DFA-based representation, we can also easily overcome this issue by applying two well-known strategies.

- The label of every chosen example, e.g., $f_{PUF}(c)$, (potentially having the metastablity situation) will be stabilized by majority voting through several oracle calls on the same example (challenge).
- A problematic example resulting in different outputs at the arbiter can be simply discarded and substituted by another randomly chosen example.

Chapter 4
PAC Learning of XOR Arbiter PUFs

> *While it is possible to attack XOR [Arbiter] PUFs using ML for small numbers of XORs, XOR [Arbiter] PUFs are widely assumed to be secure against ML attacks if enough XORs are used.*
>
> [11]

The content of this chapter is mainly based on [35]. In this chapter, besides the development of a PAC learning framework for XOR Arbiter PUFs, a theoretical limit for ML attacks as a function of the number of the chains and the number of Arbiter PUF stages has been established. Furthermore, we show that our approach deals with the noisy responses in an efficient fashion so that in this case, the maximum number of CRPs collected by the attacker is polynomial in the noise rate. Our rigorous mathematical approach matches the results of experiments, which can be found in the literature. Last but not least, on the basis of learning theory concepts, this chapter explicitly states that the current form of XOR Arbiter PUFs may not be considered as an ultimate solution to the problem of insecure Arbiter PUFs.

Overview of this chapter: Section 4.1 introduces the attacks against XOR Arbiter PUFs proposed in the literature and summarizes the contributions of our work. Section 4.2 devoted to the background information about the representation of XOR Arbiter PUFs. Section 4.3 provides details on our learning framework, which results in modeling XOR Arbiter PUFs with given levels of *accuracy* and final model delivery *confidence*. In Sect. 4.4, it is proved that the proposed algorithm can be adapted to model noisy XOR Arbiter PUFs. An exhaustive discussion about the theoretical and practical considerations is presented in Sect. 4.5.

© Springer International Publishing AG 2018
F. Ganji, *On the Learnability of Physically Unclonable Functions*, T-Labs
Series in Telecommunication Services, https://doi.org/10.1007/978-3-319-76717-8_4

4.1 Introduction

As discussed in previous chapters, aiming at mathematically *cloning* an Arbiter PUF, the attacker collects a set of challenge-response pairs (CRPs), and *attempts* to come up with a model that can *approximately* predict the response of the PUF to an arbitrarily chosen challenge. Most of the ML attacks against Arbiter PUFs benefit from the linear additive model of an Arbiter PUF. This forces a migration to modified structures of Arbiter PUFs, in which non-linear effects are added to the PUF in order to impair the effectiveness of ML attacks. To this end, XORing the responses of multiple Arbiter PUFs has been demonstrated as a promising solution [107].

However, it has been shown that more advanced ML techniques can still break the security of an XOR Arbiter PUF with a limited number of arbiter chains (hereafter called chains) [95]. Going beyond this limited number is suggested as a counter-measure by PUF manufacturers, although they have encountered serious problems, namely the increasing number of noisy responses as well as optimization of the silicon area required on the respective chip [93]. Even in this case, physical side-channel attacks, such as photonic emission analysis, can physically characterize XOR Arbiter PUFs regardless of the number of XORs [109]. In another attempt a combination of ML attacks with non-invasive side channel attacks (e.g., power, timing, and photonic emission), so-called *hybrid attacks*, is suggested to model XOR Arbiter PUFs, with the number of chains exceeding the previously established limit [11, 31, 97].

The latter attacks are cost-effective due to their non-invasive nature, and therefore, they might be preferred to the semi-invasive one in practice. However, in contrast to pure ML techniques (i.e., without any side channel information), using side channel information in combination with ML techniques requires physical access to the device and reconfiguration of the circuits on the chip, which are not always feasible in a real scenario [97]. Therefore, it is still tempting to develop new pure ML techniques to break the security of XOR Arbiter PUFs, with an arbitrary number of chains. Nevertheless, it is still unclear how many chains should be XORed to ensure the security of Arbiter PUFs against ML attacks. Moreover, when applying current ML attacks, the maximum number of CRPs required for modeling an XOR Arbiter PUF, with given levels of *accuracy* and final model delivery *confidence*, is not known.

Although in Chap. 3 we have shown how a single chain Arbiter PUF can be represented by a DFA, we claim that for the XOR Arbiter PUFs, a more compact representation can be adopted to improve the time complexity of the attack. Furthermore, to deal with noisy responses of an XOR Arbiter PUF more efficiently, an approach not relying on majority voting can be applied.

Here we present a new framework to prove to what extend XOR Arbiter PUFs can be learned in polynomial time, for given levels of accuracy and confidence. The main contributions of our framework are summarized as follows:

Finding a theoretical limit for ML techniques to learn XOR Arbiter PUFs in polynomial time. Under a well-known representation of an XOR Arbiter PUF, we provide a theoretical limit as a function of the number of Arbiter PUF stages and the number of chains, where an XOR Arbiter PUF can be provably learned in polynomial time.

Learning of an XOR Arbiter PUF for given levels of accuracy and confidence. With regard to the proposed limit, we present an algorithm, which learns the challenge-response behavior of an XOR Arbiter PUF, for given levels of accuracy and confidence. The run time of this algorithm is polynomial in the number of the Arbiter PUF stages, the number of chains, as well as the levels of accuracy and confidence. Moreover, our approach requires no side channel information.

Modeling the XOR Arbiter PUF even if the responses are noisy. A model of noise fitting the purpose of our ML framework is applied to prove that even in the presence of noise, the run time of our algorithm is still polynomial in the number of the Arbiter PUF stages, the number of chains, levels of accuracy and confidence, and the noise rate.

4.2 LTF Representation of XOR Arbiter PUFs

An XOR Arbiter PUF consists of k different chains, all with the same number of stages n. Note that XOR Arbiter PUFs that come within the scope of this work are composed of Arbiter PUFs fed with the same challenge. Therefore, schemes with different challenges fed into the Arbiter PUFs, e.g., the one proposed in [124], are beyond the scope of our work. The responses of all Arbiter PUFs are XORed together to generate the final response, see Fig. 4.1. Hence, the response of the XOR Arbiter PUF can be defined as

$$f_{XOR}(c) = \bigoplus_{j=1}^{k} f_{j\text{th Arbiter PUF}}(c).$$

The first step that should be taken in our framework is discritization of delays, as described for Arbiter PUFs as well (see Sect. 3.2.1). Following the procedure introduced in the previous chapter, in each stage (see Fig. 4.2), e.g., ith stage, $\overline{\beta}_{i,j}$ can be mapped into an integer value $\beta_{i,j}$ ($1 \leq j \leq 4$). It is known that $\overline{\beta}_{i,j} \in [\mu_i - 3\sigma_i, \mu_i + 3\sigma_i]$ with probability 99.7%. Now we define the mapping $f_{int} : \mathbb{R} \to \mathbb{Z}$ so that for all $\overline{\beta}_{i,j} \in [\mu_i - 3\sigma_i, \mu_i + 3\sigma_i]$, when applying this mapping, we have

$$\beta_{i,j} = f_{int}(\overline{\beta}_{i,j}) = \left\lceil (\overline{\beta}_{i,j} - \mu_i + 3\sigma_i)/\gamma \right\rceil.$$

Without loss of generality, we assume that $\mu_1 = \cdots = \mu_n$ and $\sigma_1 = \cdots = \sigma_n$. Consequently, the minimum and the maximum of the real valued delays $\overline{\beta}_{i,j}$ ($1 \leq i \leq n$ and $1 \leq j \leq 4$) are mapped into 0 and $m = \left\lceil \frac{6\sigma}{\gamma} \right\rceil$, respectively (for more details see [36]). Furthermore, similarly, $\overline{\Delta}$ can be mapped with a high probability to

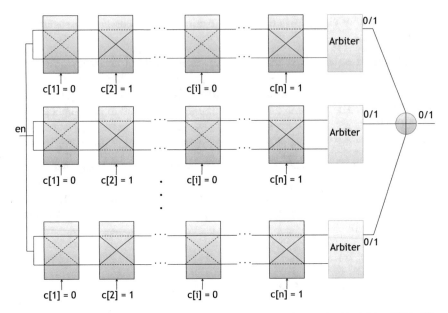

Fig. 4.1 Schematic of an XOR Arbiter PUF. It consists of k chains of n-bit Arbiter PUFs. The responses of all Arbiters are XORed together to generate the final binary response

an integer value $\Delta \in \mathbb{Z}$ lying within a finite interval. In this case, the response of the Arbiter is "1" if $\Delta > 0$, whereas it is "0" if $\Delta < 0$. The Arbiter is in the metastable condition, if $\Delta = 0$.

Here we briefly describe the LTF-based representation of an Arbiter PUF, which is widely adopted [39, 74, 95], and how such representation can be extended to establish an LTF representation of XOR Arbiter PUFs. Consider the delay vector ω^T defined as follows:

$$\omega^T = (\omega_1, \omega_2, \ldots, \omega_{n+1}) \text{ with } \begin{cases} \omega_1 = \frac{\varphi_{i,0} - \varphi_{i,1}}{2} \\ \omega_i = \frac{\varphi_{i-1,0} + \varphi_{i-1,1} + \varphi_{i,0} - \varphi_{i,1}}{2}, & 2 \leq i \leq n \\ \omega_{n+1} = \frac{\varphi_{n,0} + \varphi_{n,1}}{2}, \end{cases} \quad (4.1)$$

where the integer valued $\varphi_{i,j}$ ($1 \leq i \leq n$ and $0 \leq j \leq 1$, as shown in Fig. 4.2) are the delay differences at the output of the ith stage. With regard to the discretization process described here and in Sect. 3.2.1, it is straightforward to show that $\varphi_{i,j}$ lies within the interval $[-m, m]$, hence, ω_i ($1 \leq i \leq n + 1$) lies within the interval $[-2m, 2m]$.

Consider a challenge string represented by a vector $\mathbf{c} = (c[1], \cdots, c[n])$. The vector $\Phi = (\Phi[1], \cdots, \Phi[n], 1)$ is the encoded challenge vector, where

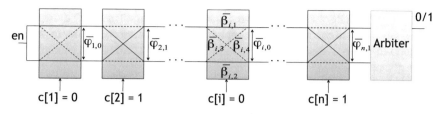

Fig. 4.2 Schematic of an Arbiter PUF composed of n so-called stages that form a chain

$$\Phi[i] = \prod_{j=i}^{n}(1 - 2c[j]).$$

According to the linear additive model of the Arbiter PUF, $\Delta = \omega^T \cdot \Phi$, cf. [61, 95]. Now let $f_{\text{map}} : \mathcal{Y} \to \{1, -1\}$, so that $f_{\text{map}}(0) = 1$ and $f_{\text{map}}(1) = -1$. The output of the Arbiter can be defined as

$$f_{PUF} = \text{sign}(\Delta) = \text{sign}(\omega^T \cdot \Phi).$$

From this equation, it is obvious that an Arbiter PUF can be represented by an $(n + 1)$-dimensional LTF. In a similar fashion, an XOR Arbiter PUF can also be represented by an LTF, where the set of final responses $(\mathcal{Y}_{XOR})$ is mapped to $\{1, -1\}$, cf. [95]:

$$f_{XOR} = \prod_{j=1}^{k} \text{sign}(\omega^T \cdot \Phi) = \text{sign}\left(\bigotimes_{j=1}^{k} \omega^T \cdot \bigotimes_{j=1}^{k} \Phi_j\right) = \text{sign}(\omega_{XOR}^T \cdot \Phi_{XOR}),$$

where $\omega_{XOR} = \otimes_{j=1}^{k}\omega_j^T$ is the tensor product of the vectors ω_j^T, and similarly $\Phi_{XOR} = \otimes_{j=1}^{k}\Phi_j$.

4.3 PAC Learning of XOR Arbiter PUFs

In this section we first present how and why an XOR Arbiter PUF can be PAC learned by the Perceptron algorithm. Furthermore, we provide the theoretical limit for the learnability of XOR Arbiter PUFs in polynomial time. Finally, our theoretical results are verified against experimental results from existing literature.

Here we show how adopting a simple transformation enables us to apply the Perceptron algorithm to PAC learn XOR Arbiter PUFs. It can be seen that for f_{XOR}, the $(n+1)^k$-dimensional vectors ω_{XOR} and Φ_{XOR} are substituted for $(n+1)$-dimensional vectors ω^T and Φ in f_{PUF}. In other words, XORing k $(n + 1)$-dimensional LTFs results in an $O((n + 1)^k)$-dimensional LTF. Therefore, if we transform the problem of learning the XOR of k $(n+1)$-dimensional LTFs into an $O((n+1)^k)$ dimensional space, this problem can be solved by applying the Perceptron algorithm. In order to support our claim, we begin with the following theorem stating that examples are linearly separable in $\mathbb{R}^{n+1}$.

Theorem 4.3.1 *For an XOR Arbiter PUF, with fixed, constant k, represented by an LTF in an $O\left((n+1)^k\right)$ dimensional space, $1/\sigma$ is polynomial in n.*

Proof For the sake of readability, the elements of ω_{XOR} are denoted by ω_i in this proof. Furthermore, elements of Φ_{XOR} are in $\{-1, 1\}$. Now we have

$$\sigma = \min_{\Phi_{XOR} \in \Phi_{XOR}} \frac{|\Phi_{XOR} \cdot \omega_{XOR}|}{\|\omega_{XOR}\| \|\Phi_{XOR}\|} = \min_{\Phi_{XOR} \in \Phi_{XOR}} \frac{|\Phi_{XOR} \cdot \omega_{XOR}|}{\sqrt{(n+1)^k \sum_{i=1}^{(n+1)^k} \omega_i^2}},$$

where Φ_{XOR} is the set of all Φ_{XOR}'s. Due to a non-trivial challenge-response behavior of the given PUF, at least one of the elements of ω must be unequal zero. Therefore, $\min(|\Phi_{XOR} \cdot \omega_{XOR}|) = 1$, and $1/\sigma = O\left((n+1)^k\right)$. ∎

Figure 4.3 illustrates how the perceptron algorithm is applied in our PAC learning framework. The learner has access to an Oracle EX, which is related to the XOR Arbiter PUF as follows:

$$EX := f_{XOR}.$$

At the first stage, the Oracle EX is called successively to collect CRPs. The maximum number of calls is denoted by r_{max}. For each CRP (e.g., $(\mathbf{c}_i, f(\mathbf{c}_i))$) a vector Φ_i is generated. Afterwards, Φ_i is transformed to Φ^i_{XOR}, which is in an $O\left((n+1)^k\right)$ dimensional space. The Perceptron algorithm predicts the response to Φ^i_{XOR}, and if its prediction and $f(\mathbf{c}_i)$ disagrees, its hypothesis will be updated.

Now we elaborate on the upper mistake bound in our framework. Following the convergence theorem of the Perceptron algorithm and Theorem 4.3.1, when $\|\Phi_{XOR}\| \leq (n+1)^{k/2}$, we have $N_{mis} = (n+1)^k/\varepsilon^2$. As an immediate corollary, we have:

Corollary 4.3.1 *Let k be constant and consider the class of XOR Arbiter PUFs over the instance space $X_n = \{0, 1\}^{(n+1)^k}$: the class of linear threshold functions such that*

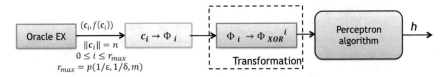

Fig. 4.3 Block diagram of the PAC learning framework applied to learn an XOR Arbiter PUF. By calling the Oracle EX at most r_{max} times, a sequence of examples is collected. Φ_i is fed into the third block corresponding to our problem transformation. The output of the third block is fed into the Perceptron algorithm

$\omega_i \in \mathbb{Z}$, and $\sum_{i=1}^{(n+1)^k} |\omega_i| = p(2m, (n+1)^k)$. Then the Perceptron-based algorithm running in time $p((n+1)^k, 4m^2, 1/\varepsilon, 1/\delta)$ can PAC learn an XOR Arbiter PUF by calling EX at most $O\left(\log(1/\delta)/\varepsilon + (4m^2(n+1))^k/\varepsilon^3\right)$ times.

There are some key implications from this corollary. First, with regard to the PAC model, the hypothesis delivered by the algorithm must be evaluable in polynomial time. The point here is that the term $(n+1)^k$, related to the Vapnik-Chervonenkis dimension of the representation [15], may grow significantly if k is not a constant. In this case, our algorithm cannot find a hypothesis in polynomial time, since the number of examples is super-polynomial in n.

Second, it is tempting that an increase in m can help to ensure the security of an XOR Arbiter PUF. The upper bound for the number of CRPs calculated according to the Corollary 2, for $\delta = 0.0001$, $k = 5$ for $n = 64$ and $n = 128$, is depicted in Fig. 4.4. This figure can provide a better understanding of the impact of m on the learnability of an XOR Arbiter PUF. A marginal increase in m cannot dramatically increase the number of CRPs required for modeling an XOR Arbiter PUF. Although an arbitrarily large m can be suggested to ensure the security of an XOR Arbiter PUF, as stated in Chap. 3 and in [36], m is restricted by technological limits, and thus cannot be arbitrarily large. In Sect. 4.5 the impact of an increase in m on the learnability will be further discussed.

Validation of the Theoretical Results

We compare our theoretical findings with the most relevant experimental results reported in [67, 95]. As reported in [67], until now no *effective* pure ML attack has been launched on XOR Arbiter PUFs with $k \geq 5$ ($n = 64$). Pure ML attacks proposed in the literature are conducted on XOR Arbiter PUFs with the maximum n being equal to 128 [95, 97]. Taking into account the long run time of pure ML algorithms, even on the powerful machines employed by [95], XOR Arbiter PUFs with $n = 256$ and $n = 512$ have been targeted only by combined modeling attacks [97]. Therefore, unfortunately, in the literature no practical limit for pure ML techniques has been reported for XOR Arbiter PUFs with $n \geq 128$.

As an attempt to compare our theoretical results to what has been observed in practice, we focus on the results reported in [95]. Note that although the algorithm applied in [95] (LR) differs from our algorithm, we can compare the number of CRPs required by them to learn an XOR Arbiter PUF with a given accuracy. The argument supporting this claim is that the hypothesis class of the LR can be "discretized" so that it becomes finite [105]. Furthermore, due to the fact that the delay values can be mapped to a finite interval of integer values (cf. Sect. 4.2), the loss function of LR is also bounded. Therefore, the LR can be converted to a PAC learning algorithm, and the maximum number of EX calls made by the algorithm is polynomial in n, k, $1/\epsilon$ and $1/\delta$. Moreover, note that the theoretical limit of learning an XOR Arbiter PUF in polynomial time is established by the Vapnik–Chervonenkis dimension of the LTF representation of an XOR Arbiter PUF, as used in [95] as well. These reasons enable us to compare their experimental results with our findings.

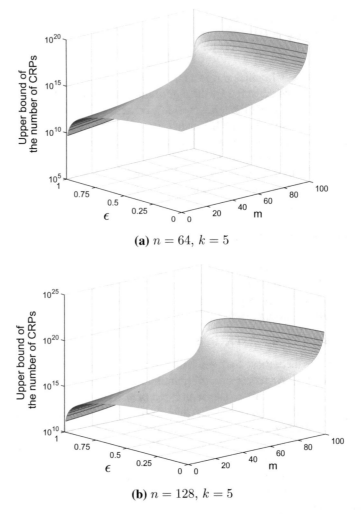

(a) $n = 64$, $k = 5$

(b) $n = 128$, $k = 5$

Fig. 4.4 Upper bound of the number of CRPs. The x-axis indicates m, whereas the y-axis shows ϵ. The z-axis corresponds to the upper bound of the number of CRPs

The authors of [95] have attempted to model 64 and 128-bit XOR Arbiter PUFs with up to 6 and 5 chains, respectively. Their results demonstrate that the proposed model can predict responses to a set of arbitrarily chosen challenges with 99% accuracy. However, the number of CRPs required for modeling a 64-bit XOR Arbiter PUF with $k = 5$ and $k = 6$ is increased drastically, comparing to those with $k \leq 4$. In this regard, the number of CRPs collected to predict the response is increased from 12000 to 80000 and 200000 for $k = 5$ and $k = 6$, respectively ($\epsilon = 0.01$). As a result, the time spent to build a model is increased from a few minutes to several hours, which shows an exponential growth. For a 128-bit XOR Arbiter PUF with $k = 5$, 500000

CRPs are required to model the XOR Arbiter PUF, with 99% accuracy, while for a PUF with $k = 4$, this number is only 24000. Consequently, the learning time is again increased exponentially.

In the above-mentioned cases, an exponential growth in the number of CRPs and the learning time can be clearly observed, when k exceeds 4. This matches the theoretical limit proposed in Sect. 4.3.

4.4 PAC Learning of Noisy XOR Arbiter PUFs

In the previous section, we have explained how the Perceptron algorithm can be applied to PAC learn an XOR Arbiter PUF. The natural and important question would be whether the proposed framework is applicable in the case of noisy responses. The term *noisy response* here refers to the response of the XOR Arbiter PUF to a challenge under either the metastable condition or the impact of environment noise. Although it has been accepted that metastability of an XOR Arbiter PUF must be solved by the PUF manufacturer, we consider this particular case for completeness. From the point of view of machine learning, this condition results in incorrect labels generated by the Oracle EX. We aim to state that an XOR Arbiter PUF can be PAC learned by applying the Perceptron algorithm, even if noisy responses are included in the collected set of CRPs.

Several versions of the Perceptron algorithm that can tolerate noise, i.e., incorrect labels of examples, have been developed (for a comprehensive survey see [57]). Here we follow the work presented in [6] to demonstrate that the original Perceptron algorithm can be further applied in the case of noisy responses. In this case, the number of CRPs required to be collected is polynomial in the number of noisy responses.

At the first stage, we define a simple but effective model of noisy Oracle EX_η [6]. In our model, the examples are drawn with respect to the relevant distribution D, and the label of each example is chosen in an independent random fashion. More specifically, after drawing an example, an unfair coin (head with probability $1 - \eta$) is flipped. If the outcome is head, the correct label is provided, otherwise the label is incorrect. It is clear that $\eta < 1/2$, since $\eta = 1/2$ means that no relevant information is provided by EX_η, and the case of $\eta > 1/2$ is irrelevant. We assume that an upper bound on η, denoted by η_b, is known. Even if this assumption may not hold in practice, following the procedure defined in [6], η_b can be estimated. It has been shown that the sample size is increased very slightly in the case of unknown η_b (for further information and the proof see [6]).

According to the Convergence Theorem of the Perceptron algorithm (see Theorem 2.4.1) in the case of noisy responses the condition $f(\Phi_i)(\mathbf{u}.\Phi_i) \geq 0$ cannot always be met. This condition relates the accuracy of the prediction performed by the Perceptron algorithm to the labels provided to that. In the case of noisy examples, the abovementioned condition should be modified so that it reflects the accuracy of the Perceptron algorithm in the presence of noise. Suppose that an example, i.e.,

$(\Phi_i, f(\Phi_i))$, is provided by EX_η. The probability that this example disagrees with any hypothesis $\mathbf{u}$ can be calculated as following:

$$\Pr[f(\Phi_i)(\mathbf{u} \cdot \Phi_i) < 0] \leq (1 - \eta)\varepsilon + \eta(1 - \varepsilon) < \eta_b + \varepsilon(1 - 2\eta_b).$$

From this inequality, it can be inferred that the expected rate of disagreement is at least η for the ideal hypothesis $\mathbf{u}$. Therefore, the separation factor of at least $\varepsilon(1 - 2\eta)$ should be between an ideal hypothesis and an approximation of that cf. [6]. As stated in the following theorem, the maximum number of mistakes that can be made by the Perceptron algorithm is polynomial in this separation.

Theorem 4.4.1 *Consider r labeled examples which are fed into the Perceptron algorithm, and let $\|\Phi_i\| \leq R$. In the case of noisy labels, let $\mathbf{u}_n$ be the solution vector with $\|\mathbf{u}_n\| = 1$, and $f(\Phi_i)(\mathbf{u}_n.\Phi_i) \geq \varepsilon(1 - 2\eta_b) > 0$. Then*

$$N_{mis} = \left(\frac{R}{\varepsilon(1 - 2\eta_b)} \right)^2.$$

It is straightforward to prove this theorem, and for more details the reader is referred to [6].

Theorem 4.4.2 *When PAC learning the noisy XOR Arbiter PUF, the maximum number of mistakes that the Perceptron algorithm can make is $N_{mis} = (n + 1)^k / (\varepsilon^2(1 - 2\eta_b)^2)$. Furthermore, the maximum number of CRPs required for PAC learning a noisy XOR Arbiter PUF is $O\left(\log(1/\delta)/(\varepsilon(1 - 2\eta_b)) + (4m^2(n + 1))^k / (\varepsilon^3(1 - 2\eta_b)^3) \right)$.*

Proof Following Corollary 4.3.1 and Theorem 4.4.1, this can be easily shown.   ■

The most important message is that this maximum number of CRPs is polynomial in n, ε, δ as well as the upper bound of η. According to experimental results when the noise rate is 2%, the number of CRPs required to learn a 128-bit XOR Arbiter PUF ($k = 4$) is approximately increased by the factor 2, in comparison to the noiseless scenario with approximately the same ϵ [95]. For the same XOR Arbiter PUF, increasing the noise rate to 5 and 10%, the number of CRPs is increased 2 and 8 times, comparing with the case of $\eta_b = 0.02$. It has been concluded that the number of noisy CRPs collected to model the XOR Arbiter PUF is polynomial in the noise rate [95], which agrees with our theoretical result.

4.5   Discussion

4.5.1   *Theoretical Considerations*

By providing the proof of vulnerability of XOR Arbiter PUFs to PAC learning we have demonstrated how fragile the security of this kind PUF can be. The concept of

PAC learning of XOR Arbiter PUFs has been already demonstrated by Hammouri et al. [45]. Although the authors benefit from an adequate representation of the XOR Arbiter PUFs, which is LTF-based [39], they could not prove the PAC learnability of the XOR Arbiter PUFs. As the Vapnik–Chervonenkis dimension of an LTF representing an Arbiter PUF is equal to $n + 1$, this family of PUF primitives is subject to PAC learning attacks [36]. It is straightforward to further prove that the Vapnik–Chervonenkis dimension of the LTF representing an XOR Arbiter PUF is $(n + 1)^k$. Therefore, for constant k an XOR Arbiter PUF with k chains (each with n stages) is also PAC learnable since the Vapnik–Chervonenkis dimension of the LTF-based representation of such PUF is finite. In this chapter, instead of sticking to this obvious fact, to PAC learn an XOR Arbiter PUF–even in the case of noisy response– we apply the Perceptron algorithm, whose run time is polynomial in n, $1/\varepsilon$, $1/\delta$, and k.

Another important aspect of our framework is the representation of an XOR Arbiter PUF. As mentioned earlier, it is clear that according to what has been observed in Chap. 3 and [36], an XOR Arbiter PUF can also be represented by a DFA with $O(n^k M^{2k})$ states. Therefore, the algorithm learning this DFA (see Algorithm 2) makes

$$O\left((1 + 2/\varepsilon \ln(1/\delta))n^k M^{2k} + 2/\varepsilon n^{2k} M^{4k}\right)$$

calls to EX. Comparing this number of calls with the number of calls that Perceptron algorithm makes to EX (see Corollary 2), it is clear that the numbers of calls made by both algorithms are polynomial in n, m, $1/\varepsilon$, and $1/\delta$. However, the algorithm presented in this chapter outperforms in terms of the number of calls, and consequently its time complexity.

Of crucial importance for our framework is how the algorithm deals with noisy responses. In this chapter, we have proposed a model of noise, which is well-studied in the PAC learning related literature and agrees with what can be seen in practice. Towards launching an ML attack, the adversary applies a set of challenges and collects the responses, where the latter might be noisy. In the literature, majority voting is suggested as a solution to deal with noisy responses [73, 95]. This can impair the performance of the proposed learning algorithm, when the attacker can observe each CRP only once and cannot do majority voting. It is even suggested that in order to reduce the effectiveness of ML attacks, the noise rate that can be observed by an attacker can be artificially increased, while the verifier still observes only a small noise rate [127]. In this latter scenario, the majority voting cannot be helpful. On contrary, we have proved that even noisy XOR Arbiter PUFs can be PAC learned without the need for majority voting.

We have stated that the maximum number of mistakes that the Perceptron algorithm can make, and consequently, the maximum number of CRPs required for PAC learning is polynomial in n, k, $1/\varepsilon$, $1/\delta$ as well as $1/(1 - 2\eta)$ in the case of noisy responses. Since we have mainly aimed to prove this, the maximum number of CRPs calculated in Sect. 4.3 ensures that the algorithm delivers an approximation of the desired hyperplane, with the probability at least $1 - \delta$. The proposed upper bound of the number of CRPs can be improved to even reduce the number of required CRPs (see for instance [15]).

4.5.2   Practical Considerations

When proving that the Perceptron algorithm can be applied to PAC learn an XOR Arbiter PUF, we take the advantage of the lessons learnt from practice, namely

- the delay values can be mapped to a finite interval of integer values, and
- the number of chains contained in an XOR Arbiter PUF (k) may not exceed a certain value.

The importance of the first fact can be recognized in the proof of the PAC learnability of an XOR Arbiter PUF (see Corollary 2). The second fact confirms that the Vapnik–Chervonenkis dimension of the LTF representing an XOR Arbiter PUF is finite. Whereas the first fact has been already reflected in Chap. 3 and [36], the second one has been only partially discussed in the literature.

The results of experiments demonstrate that XORing more than a certain number of chains can be challenging in the practice [93]. In the experiments conducted by Rostami et al. [93], different XOR Arbiter PUFs designed on 10 Xilinx Virtex 5 (LX110) FPGAs at the nominal condition (temperature = 35 °C and VDD = 1 V) are employed. For $k = 4$, it is reported that the noise rate is $\eta = 23.2\%$, and a change in the condition (e.g., reducing VDD) may result in an increase in the noise rate up to 43.2%. They have suggested that the noise rate of an XOR Arbiter PUF can be approximated by summing the noise rate of each individual Arbiter PUF. In an attempt to precisely calculate the noise rate of an XOR Arbiter PUF, the binomial distribution of the error has been taken into account [124, 127]. With regard to this noise analysis, assuming that the noise rate is about 4% for an Arbiter PUF designed on 65 nm technology,[1] the noise rate of the XOR Arbiter PUF with $k = 12$ is about 33%. Note that under the condition that the noise ratio would be approximately 50%, even majority voting cannot be helpful so that the PUF cannot be verified. Although the analyses conducted in [124, 127] have shown that the noise rate may not exceed 50%, when n is large and $k \gg (\ln n)$, concerns about PUF implementations cannot be ignored. For instance, an important factor limiting k is the silicon area used for constructing an XOR Arbiter PUF. Based on a rough estimation reported in [67], the silicon area required for constructing an XOR Arbiter PUF with k chains is k times larger than a single Arbiter PUF.

Despite the implementation and technological limits on k, we have proved theoretical limits on when an XOR Arbiter PUF can be learned in polynomial time. In the literature, it has not so far been stated how the learnability is theoretically limited. Nevertheless, the empirical upper bound reported in [67] and the experimental results in [95] are in line with our theoretical limit. Moreover, the experimental results presented in [114] provide evidence that supports our findings. It has been shown that when $n = 64$ and $k \geq 4$, the number of CRPs required for the ML attack, and consequently the time complexity, is increased drastically. The same observation is repeated for $n = 128$ and $k \geq 5$. These emphasize the importance of our approach,

---

[1]For an Arbiter PUF designed on 65 nm technology, a typical value of the noise rate is about 4% [93].

in which not only the limit of the learnability in polynomial time is identified but also no side channel information is required to PAC learn the XOR Arbiter PUFs under this limit. With respect to this theoretical limit, to evaluate the security of an XOR Arbiter PUF the following scenarios can be distinguished:

- n is small (e.g., $n \leq 32$): in this case, the security can be easily broken by adopting a brute-force strategy.
- n is large (i.e., no brute-force strategy is applicable) and $k \gg (\ln n)$: under this condition, the XOR Arbiter PUF cannot be learned in polynomial time. However, a practical implementation of such an XOR Arbiter PUF can be challenging due to the technological limits, more specifically the noisy responses and the silicon area.
- n is large and $k \ll (\ln n)$: the XOR Arbiter PUF can be PAC learned.

Note that in the second scenario, the infeasibility of applying a PAC learning framework does not rule out the possibility that empirical ML or hybrid attacks can be successfully launched cf. [11, 31, 97]. In the latter scenario, it can be thought that an increase in m may lead to a more secure XOR Arbiter PUF. From a theoretical point of view, more CRPs are required for PAC learning an XOR Arbiter PUF with large m, the number of CRPs is still polynomial in m, n, constant k and levels of accuracy and confidence. On the other hand, from a practical perspective, a chip designed with the large σ might not work properly. Moreover, it can be suggested to produce Arbiters with high precision in order to enlarge m. In this case, the cost of the chip is increased dramatically.

In previous studies, e.g., [95, 97], powerful and costly machines have been employed to prove the concept of learnability of XOR Arbiter PUFs. It might not be convenient to run a ML algorithm on such machines, particularly for XOR Arbiter PUFs with large k and n. Since our concrete proofs state how the security of XOR Arbiter PUFs can be broken in polynomial time, it seems redundant to conduct a simulation or an experiment concerning this issue. Last but not least, we emphasize that protocols relying on the security of XOR Arbiter PUFs cannot be considered as an ultimate solution to the issue of insecure Arbiter PUFs. As it has been also stated in [22], none of the XOR Arbiter PUF-based protocols in its current form can be thought of as being perfectly secure.

Chapter 5
PAC Learning of Ring Oscillator PUFs

> *One of the features of modeling PUF circuits through ML*
> *techniques is that except for Arbiter PUFs, other PUFs cannot*
> *be modeled very satisfactorily in a way to suggest which ML to*
> *apply to model them. [...] This observation suggests heuristic*
> *techniques which are effective in estimating input-output*
> *relationships [...].*

[98]

This chapter covers the principles of a PAC learning framework applied to ring oscillator (RO) PUFs, which have been first introduced in [34]. Parts of this paper have been slightly adapted to be involved in this thesis.

This chapter demonstrates that inherent characteristics of RO-PUFs, although being *hidden* from an adversary, account for the success of the previously proposed (heuristic) attacks and is the key factor in the success of our framework as well. These characteristics are helpful to establish a polynomial-sized representation of RO-PUFs, and consequently, provably learn an RO-PUF for given levels of accuracy and confidence. In addition, when comparing the number of CRPs required for our attack with already existing bounds calculated by applying heuristic techniques, our proposed bound is provably better. Finally, by conducting experiments, we complement the proof provided in our PAC learning framework.

Overview of this chapter: Similar to previous chapters, we begin with a brief literature review in Sect. 5.1. Afterwards, a new representation of RO-PUFs is established in Sect. 5.2, which enables us to PAC learn these primitives (Sect. 5.3). Our practical results and a discussion of them are presented in Sect. 5.4.

© Springer International Publishing AG 2018
F. Ganji, *On the Learnability of Physically Unclonable Functions*, T-Labs
Series in Telecommunication Services, https://doi.org/10.1007/978-3-319-76717-8_5

5.1 Introduction

In conjunction with Arbiter PUFs, RO PUFs are classified as delay-based PUFs that attract attention thanks to their easy and inexpensive implementations on different platforms [67, 107]. Arbiter PUFs and RO-PUFs share the common feature of using the various propagation delays of identical electrical paths on the chip to generate a *virtually* unique output. Nevertheless, coming under several attacks, it has been shown that the security of these PUFs can be comprised, and therefore, the *unclonability* and *unpredictability* features promised by the manufacturer are not absolutely supported.

Similar to Arbiter PUFs, it has been stated that RO-PUFs are susceptible to semi-invasive side channel analysis [78]. Moreover, as an instance of modeling attacks against RO-PUFs, Rührmair et al. have applied Quicksort algorithm to model an RO-PUF in the case that the adversary cannot control the challenges, although the challenges can be eavesdropped [95]. As another example, a new attack on RO-PUFs has been introduced, whose key success factor is the availability of the helper data used to compensate the impact of the noise on the responses [83]. Obviously, in the absence of the helper data, which is a likely scenario in practice, their attack cannot succeed. In another attempt to develop an ML method that can be applied to compromise the security of an RO-PUF, a genetic programming approach has been employed [98]. Although their results are promising regarding the prediction accuracy of the obtained models, neither the scalability of the approach nor the probability of delivering the final model has been discussed. These concerns have been already addressed for Arbiter and XOR Arbiter PUFs [35, 36] (see Chaps. 3–4). Unfortunately, such a thorough analysis has not been developed for RO-PUFs so far, and solely empirical modeling attacks have been suggested. The lack of such precise analyses further supports the development of ad hoc solutions to their security problems. Therefore, several different implementation methods have been proposed in order to improve the security of RO-PUFs [71, 81], despite the fact that this primitive is inherently vulnerable to ML attacks. Therefore, goals of this chapter are:

Establishing a fit-for-purpose representation of RO-PUFs. In addition to the feature of being polynomial-sized, our proposed representation can be easily built by collecting CRPs. Therefore, when comparing to other complicated and sophisticated representations, it can be rapidly established. Due to this representation, we propose an algorithm that can learn an RO-PUF for given levels of accuracy and confidence.

Mathematical proof of the vulnerability of RO-PUFs to our ML attack. The number of CRPs required to launch our attack is carefully calculated. We prove that this small number is indeed polynomial in the number of ring oscillators.

Providing a proof of concept of how our attack performs in practice. By conducting experiments, we evaluate the effectiveness of our attack.

5.2 DL Representation of RO-PUFs

The manufacturing variations in the delays of circuit gates have been used to design an RO-PUF [107]. The first architecture proposed by Suh et al. is composed of N identically designed oscillator rings, whose frequencies are compared pairwise to generate the binary output of the RO-PUF, see Fig. 5.1. Although $N(N-1)/2$ pairs of oscillators are possible for this architecture, due to the particular ascending order of the frequencies, the number of responses cannot exceed $\log_2(N!)$ [50]. Consider the scheme of an RO-PUF depicted in Fig. 5.1 that features N ring oscillators. By applying a binary challenges, the frequencies of two ring oscillators are compared to generate the response. Clearly, the challenges applied to two multiplexers should not be identical; otherwise, only N different pairs of ring oscillators can be selected, and consequently, solely N responses can be obtained. Hence, more formally, the k- bit challenge $c_1 c_2 \cdots c_k$ applied to one of the multiplexers (e.g., the upper one) to select the first ring oscillator, whereas the k- bit challenge $c'_1 c'_2 \cdots c'_k$ is fed into the second multiplexer to select the second ring oscillator. Note that c'_i is not the complement of c_i, but being chosen randomly and independently from c_i. As required by our approach, without loss of generality, we denote the appropriate challenge applied to the RO-PUF by the binary string $c_1 c_2 \cdots c_k c'_1 c'_2 \cdots c'_k$, where $k = \log_2 N$ and $c_i \neq c'_i$ ($1 \leq i \leq k$). It may be thought that the number of bits in a challenge can represent a measure of the security of this type of PUFs. However, in practice, the number of ring oscillators implemented on a chip is a more limiting, and a determinant factor. The influence of this factor on the uniqueness of the RO-PUF and the silicon area footprint has been discussed in the literature, e.g., [71].

The vulnerability of RO-PUFs to modeling attacks has so far been revealed in the literature cf. [95]. When launching a modeling attack on these PUFs, two scenarios can be considered.

- In the first scenario, the attacker can *selectively* apply the desired challenges to figure out the ascending order of frequencies of the rings. In this scenario, the number of CRPs can be $O(N \log_2 N)$, or in an extreme case, $N^2/2$, where the attacker collects all the possible CRPs. Of course it is possible in theory, but in practice the attacker may not have direct access to the challenges and, consequently, this types of attacks may fail.
- In an advance and a more realistic scenario that is considered in our work, the attacker can solely collect the challenges randomly applied to the PUF and the respective responses.

In order to establish a proper representation of an RO-PUF, we focus on a widely accepted implementation of RO-PUFs as proposed in [107], see Fig. 5.1. Nevertheless, modified architectures have been proposed in [122, 125, 126]. Potential attacks against them are discussed briefly in Sect. 5.3.

Consider an RO-PUF that features N ring oscillators. The challenge is a $2k$-bit binary string $c_1 c_2 \cdots c_k c'_1 c'_2 \cdots c'_k$ as discussed in Sect. 5.2. When applying this binary challenge to an RO-PUF, the string $c_1 c_2 \cdots c_k$ determines the first ring oscillator to be

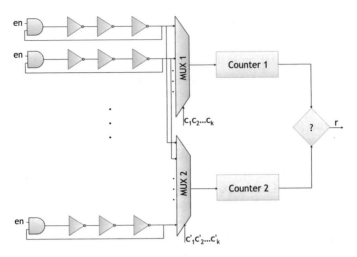

Fig. 5.1 An RO-PUF with N ring-oscillators. By applying challenges to two multiplexers, two ring-oscillators are selected and their outputs are connected to the clock inputs of 2 counters. The counters count the number of the rising edges during a predefined time period. Finally, the state of the counters are compared by the comparator placed at the end of the PUF to generate a binary response

selected, whereas the string $c_1'c_2'\cdots c_k'$ determines the second one. By comparing the frequencies of these ring oscillators, the final response of the RO-PUF is generated. We define the binary to one-hot encoded mapping $f_{map} : \{0, 1\}^k \rightarrow \{0, 1\}^N$ that maps a binary string, e.g., $c_1 c_2 \cdots c_k$, to a one-hot string $x_1 x_2 \cdots x_N$. Therefore, all Boolean attributes of the mapped string are "0", except solely one of them, e.g., the jth attribute, that is "1" corresponding to the selected ring oscillator.

By performing the mapping f_{map} on each challenge, we obtain two one-hot encoded strings merged to a single mapped challenge $x_1 x_2 \cdots x_N$. In other words, if $f_{map}(c_1 c_2 \cdots c_k) = x_1 x_2 \cdots x_N$, where $x_i = 1$ and for the second string we obtain $f_{map}(c_1'c_2'\cdots c_k') = x_1'x_2'\cdots x_N'$, where $x_j = 1$, we can merge the mapped strings to a single string $x_1 x_2 \cdots x_N$, where x_i and x_j are "1". This step can be performed easily by, e.g., adding the respective attributes of two strings together. Let the set $X_N = \{0, 1\}^N$ denote the set of all mapped challenges. Note that according to the definition of f_{map}, only two non-zero Boolean attributes of each mapped challenge are drawn from V_N.

Similar to the other types of PUFs, an RO-PUF can be represented by the function $f_{RO} : X_N \rightarrow Y$, where $Y = \{0, 1\}$, and $f_{RO}(x_1 x_2 \cdots x_N) = y$. Obviously, this mapping represents a Boolean function. More precisely, we define each mapped challenge as being a term (e.g., f_i) so that $f_i \in C_2^N$. Now the list L containing r CRPs represents a $2 - DL$.

Algorithm 3 The algorithm for PAC learning of a k-DL as proposed by [90]

Require: The set S containing r pairs $(f_1, v_1), \cdots, (f_r, v_r)$, where $(1 \leq r \leq m)$
Ensure: L that is a k-DL

1: $T := \emptyset$
2: $j = 1$
3: **while** S is not empty **do**
4: Find a term t in $M_{n,k}$ so that all f_i $(1 \leq i \leq r)$ make t true, and their corresponding v_i are either "0" or "1"
5: $T \leftarrow t$
6: **if** t corresponds to positive examples **then**:
7: $v = 1$
8: **else**
9: $v = 0$
10: **fi**
11: **return** (t, v) as the jth item of L
12: $S := S - T$
13: $j = j + 1$
14: **od**

5.3 PAC Learnability of the 2 − *DL* Representing the RO-PUF

In order to prove that RO-PUFs are indeed PAC learnable under the DL representation, we follow the procedure introduced in [90]. We first prove that a $2 − DL$ representing an RO-PUF has a polynomial size.

Theorem 5.3.1 *A 2 − DL representing an RO-PUF is polynomial-sized.*

Proof The maximum number of elements in a k-DL, in the general case, has been determined in [90], and it has been shown that

$$size(k − DL) = O\left(\log_2\left(3^{|M_{n,k}|}(|M_{n,k}|)!\right)\right),$$

where $|M_{n,k}|$ is the cardinality of the set of the monomials $M_{n,k}$. This can be proved since, in the DL, each term from $M_{n,k}$ can be labeled by "0", "1", or "missing". Furthermore, no order for the elements of the list is defined.

Now we put emphasis on the size of the DL representing an RO-PUF. Obviously, according to our particular definition of the strings $x_1 x_2 \ldots x_N$, in our DL the maximum number of possibles elements is $N(N − 1)/2$. However, according to the ascending order of the frequencies of the ring oscillators, a list containing $O(N − 1)$ terms is completely expressive. This can be easily understood due to the relationship between the Boolean attribute x_i and the frequency of the ith ring oscillator. Therefore, the size of our $2 − DL$ in bits (i.e., the size of the representation) is $O\left((N − 1)\log_2(N − 1)\right)$. ∎

To complete our proof of the RO-PUF learnability, in addition to Theorem 5.3.1, we have to provide a polynomial-time algorithm that can generate a DL, when being fed by a set of labeled examples (so-called *sample*). To this end, we apply a classical polynomial-time algorithm that learns a k-DL ($k = 2$ in our case) for given levels of accuracy and confidence, when it is given a polynomial number of examples [90]. For the sake of completeness, the main steps of such algorithm proposed by Rivest have been presented in Algorithm 1.1 (see [90] for more details on the algorithm).

Furthermore, to PAC learn an RO-PUF, the upper bound of the number of CRPs required by Algorithm 1.1 can be calculated according to the polynomial learnability theorem proved by Blumer et al. [14].[1]

Theorem 5.3.2 *Assume that the learner is given access to Oracle $EX := f_{RO}$, and can call it successively to collect t independently drawn examples (i.e., CRPs). To PAC learn the RO-PUF for given ε and δ, under the $2 - DL$ representation, the number of CRPs required to be collected is bounded by*

$$t = O\left(\frac{1}{\varepsilon}\Big((N-1)\log_2(N-1) + \log_2\Big(\frac{1}{\delta}\Big)\Big)\right).$$

As pointed out in Sect. 5.2, the number of ring oscillators heavily affects the uniqueness and the silicon area footprint of RO-PUFs. In other words, although $N = 2^k$ and k can be increased in theory, N and consequently k cannot be arbitrarily increased due to the restrictions imposed by the technological properties of ICs. Hence, not only in our approach but also in previously proposed attacks (e.g., [95]) the number of CRPs required to characterize the challenge-response behavior of RO-PUFs is presented as a function of N, i.e., the number of ring oscillators. An important message conveyed by Theorem 5.3.2 is that the maximum number of CRPs needed to be collected by the attacker is polynomial in N, and more importantly, it is asymptotically better than the bound estimated in [95].

To give a better understanding of the impact of a change in ε and δ on the number of CRPs, the upper bound of the number of CRPs calculated according to Theorem 5.3.2 is depicted in Fig. 5.2. The curve is drawn for $N = 1024$ and different ε and δ values.

PAC Learnability of the Self-decision RO-PUF

The core idea behind the design of this type of PUFs [125], is that combination of frequencies of the ring oscillators forming the PUF can be a countermeasure against ML attack. Although this assumption was correct at the time, when this design has been proposed, the invalidity of that can be proved with regard to the results presented in Chaps. 3–4 and [35, 36]. The architecture proposed in [125, 126] is similar to the architecture of Arbiter PUFs. However, the main difference is that the frequencies of the ring oscillators (instead of delays of stages in an Arbiter PUF) are added together. Following the procedure proposed in Chap. 3 and [36], these real-valued frequencies of ring oscillators can be mapped to a limited integer interval. This enables us to construct a deterministic finite automaton (DFA) representing the sum RO-PUF, and then PAC learn it.

A more interesting design suggested by Yu et al. [122] relies on the fact that more complex recombination functions, e.g., for RO-PUFs, XOR function can provide

---

[1]We refer the reader to [14] for the proof of Blumer's theorem.

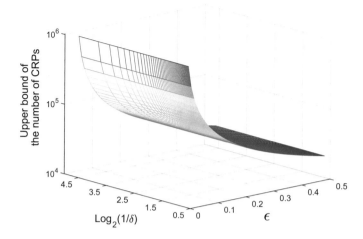

Fig. 5.2 Upper bound of the number of CRPs required for PAC learning of an RO-PUF with 1024 ring oscillators

additional robustness against ML attacks. The proposed architecture shares several similarities with XOR Arbiter PUFs, and in a similar fashion can be represented by LTFs [7]. Although in general the Vapnik–Chervonenkis dimension of these PUFs can be exponential in the number of ring oscillators (in the case of RO-PUFs), when this number does not exceed the upper bound $\ln N$, the RO-PUF is indeed PAC learnable (for the proof see Chap. 4 and [35]). On the other hand, when the number of ring oscillators exceeds this upper bound, hybrid attacks similar to what has been proposed in [31, 110] can be applied to break the security of the RO-PUF.

5.4 Results and Discussion

In this section we provide simulation results to validate our theoretical findings. To this end, one can adopt the results of large-scale experiments reported in [69]. In addition to these results, further measurement results are publicly accessible in a dataset [102]. In this dataset the measurement results containing 100 samples of the frequency of each and every ring oscillators of RO-PUFs are collected. Each RO-PUF is composed of 512 ring oscillators implemented on 193 90-nm Xilinx Spartan (XC3S500E) FPGAs. Since we aim to evaluate the effectiveness of our attack against RO-PUFs with a different number of the ring oscillators, we develop an RO-PUF simulator, whose inputs are frequencies of the ring oscillators. First, our simulator randomly selects N ($N = 128, 256, 512, 1024$) frequencies associated with N different ring oscillators. Afterwards in order to create a set of CRPs, random challenges are applied to the PUF to select a pair of ring oscillators. The indexes of selected ring oscillators and their corresponding responses were stored in a dataset to be learned by the ML algorithm proposed in Sect. 5.3.

To learn the CRPs under the DL representation, we have used the open source ML software Weka [44], providing a firm platform for conducting experiments. In our experiments 10 GB RAM of our machine is used. Moreover, the physical core of the machine is an Intel Core 2 Duo (Penryn) running at 2.4 GHz. Experiments conducted in Weka consist of two phases, namely the training and the validation phases. The examples fed into an algorithm can be divided into equally sized subsets (so-called folds) to perform cross-validation. For instance, when 10 equally sized subsets are fed into an algorithm, the model is established based on 9 subsets, and then the obtained model is validated on the remaining subset. This process is repeated 10 times so that each subset is used once as the validation dataset and 9 times as the training dataset. Finally, the results obtained for all 10 experiments are averaged to generate a single model. With respect to our setting, this 10 fold cross-validation method is applied in order to evaluate the error of the obtained model (ε). Since the model is always delivered in our experiments, it can be interpreted that δ is very close to zero in our case, as pre-defined and have coded in Weka.

The results of the experiments for several different RO-PUFs have been depicted in Fig. 5.3. As expected, for the same number of CRPs, the error of the model is higher for RO-PUFs with the higher number of ring oscillators. Furthermore, in our experiments, we increase the number of CRPs fed into the algorithm to the extent that a model with a sufficiently small error is obtained. Nevertheless, the maximum number of CRPs given to the algorithm is less than the upper bound calculated in Sect. 5.3. It can be seen in Fig. 5.3a that for each RO-PUF the error is significantly reduced, when increasing the number of CRPs collected to launch the attack. The maximum time taken to deliver the model of the RO-PUFs, corresponding to the

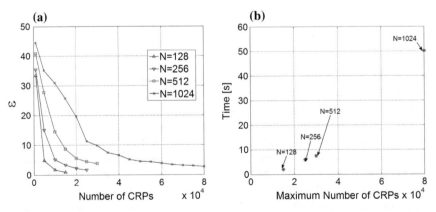

Fig. 5.3 **a** The number of CRPs required to PAC learn RO-PUFs with different numbers of ring oscillators. Clearly, when increasing the number of CRPs collected for the attack, the error of the obtained model is reduced. **b** Time taken to deliver the model, if the algorithm is fed by the maximum number of CRPs for each RO-PUFs

maximum number of CRPs given to the algorithm, is presented in Fig. 5.3b. The time complexity is increased for RO-PUFs with a higher number of ring oscillators. However, it is still polynomial in the number of the ring oscillators.

In an attempt to compare our theoretical and practical findings with results previously reported in the literature, we take into consideration the results reported in [95] and [98]. We consider the worst-case scenario from the adversary perspective, where she can only eavesdrop the CRPs and cannot apply any desired challenges. The crucial difference between the algorithm proposed in [95] and our work is that the delivery of the model is not guaranteed in their framework. Furthermore, although for different RO-PUFs virtually the same numbers of CRPs are suggested in our work and [95], the number of CRPs required for launching their attack is estimated empirically, and is heavily depending on their limited number of experiments. On the contrary, the upper-bound of the number of CRPs required to launch our PAC learning attack is calculated precisely. More importantly, the upper bound calculated with regard to our framework is asymptotically better than their estimated bound.

In a similar fashion, in spite of the fact that the algorithm proposed in [98] might deliver a kind of Boolean function for a given RO-PUF, neither the delivery of the model is ensured nor the scalability of the algorithm is discussed. The experiments conducted to evaluate the feasibility of their attack are performed only on RO-PUFs with 128 ring oscillators. Moreover, when increasing the number of CRPs to achieve a more accurate model, the time taken to generate their proposed model is increased drastically. Unfortunately, the time complexity of their model has not been further discussed.

Finally, from our results as well as what has been reported in the literature, we can draw the conclude that similar to other PUF families studied in this work, RO-PUFs cannot be considered secure regarding their current schemes. We put emphasis on the fact that we provide a proof for the learnability of RO-PUFs, which has not been addressed before in the literature.

Chapter 6
PAC Learning of Bistable Ring PUFs

> *The BR-PUF was proposed as a promising circuit-based strong PUF candidate, given that a simple model for its behavior is unknown by now and hence modeling-based attacks would be hard.*

[101]

This chapter presents the foundation of our PAC learning approach applied to bistable ring PUFs (BR-PUFs). The idea of applying this approach has been first published in [32, 33]. Some sections of this paper have been slightly reworked to be included in this thesis.

In this chapter, we present a key result stating that in general the responses of PUF families are not equally determined by each and every bit of their respective challenges, but only a subset of the bit positions called influential bits. Furthermore, to the best of our knowledge, we for the first time suggest applying Boolean and Fourier analyses in order to assess the security of PUFs. In this regard, we introduce new metrics and notions such as the *average Sensitivity of Boolean Functions* that are well-known and widely used in ML theory and may provide special insights into the physical design of secure PUFs in the future. Relying on this basis, we deal with the issue of strong PAC learning of the challenge-response behavior of PUFs, whose functionality lacks a precise mathematical model, e.g., BR-PUF family. Moreover, in addition to the results of the Boolean analysis, the feasibility and effectiveness of our machine learning attack have been evaluated by conducting experiments on the real-world BR-PUFs.

Overview of this chapter: The structure of this chapter differs from the other chapters since due to the unknown mathematical model describing the internal functionality of the BR-PUF family, a much more sophisticated ML attack should be mounted on them. Hence, we first discuss the architecture of these PUFs (see Sect. 6.2), after a brief introduction (Sect. 6.1). Afterwards, in Sect. 6.3, we shift our focus to the

© Springer International Publishing AG 2018

F. Ganji, *On the Learnability of Physically Unclonable Functions*, T-Labs
Series in Telecommunication Services, https://doi.org/10.1007/978-3-319-76717-8_6

Boolean analysis that helps us to come up with measures to PAC learn the BR-PUF family (Sect. 6.4). The experimental results and a discussion of them are presented in Sect. 6.5.

6.1 Introduction

In the previous chapters, we have proposed the PAC learning framework applied against Arbiter, XOR Arbiter, and RO-PUFs. One of the key ideas behind our approaches is that knowing about the mathematical model of the PUF functionality enables the adversary to establish a proper *hypothesis representation* (i.e., mathematical model of the PUF), and then it is possible to PAC learn this representation. This gives rise to the question of whether a PUF can be PAC learned without prior knowledge of a precise mathematical model of the PUF.

BR-PUFs [20] and twisted BR-PUFs (TBR-PUF) [101] are examples of PUFs, whose functionality cannot be easily translated to a precise mathematical model. In an attempt, the authors of [101, 120] suggested simplified mathematical models for BR-PUFs and TBR-PUFs. However, their models neither precisely reflect the physical behavior of these architectures nor the accuracy of their model can be enhanced up to a certain, desired level. From the point of view of machine learning theory, the latter can be explained by the fact that the model applied to represent these PUFs has not been chosen entirely correct [56].

This chapter presents a sound mathematical machine learning framework, which enables us to PAC learn the BR-PUF family (i.e., including BR- and TBR-PUFs) without knowing their precise mathematical model. In this regard, the most remarkable results are as follows:

Strong ML attacks against PUFs without available mathematical model. We prove that even in a worst case scenario, where the internal functionality of the BR-PUF family cannot be mathematically modeled, the challenge-response behavior of these PUFs can be PAC learned for given levels of accuracy and confidence.

Evaluation of the applicability of our framework in practice. In order to evaluate the effectiveness of our theoretical framework, we conduct extensive experiments on BR-PUFs and TBR-PUFs, implemented on a commonly used FPGA.

In addition, our framework contributes to the following aspect related to the security assessment of PUFs in general:

Exploring the inherent mathematical properties of PUFs. One of the most natural and commonly accepted mathematical representations of a PUF is a Boolean function. This representation plays a major role in investigating properties of PUFs, which are observed in practice, although they have not been precisely and mathematically described. One of the widely accepted properties of PUFs, exhaustively studied in our work, is related to the *silent* assumption that each and every bit of a challenge has equal influence on the respective response of a PUF. We prove that this assumption is not valid for all PUFs. This phenomenon has been already observed in practice and is most often attributed to implementation imperfections. Our rigorous mathematical proof states the existence of influential bit positions, which holds for PUFs.

6.2 Architecture of the BR-PUF Family

In this section, we explain the architectures of BR- and TBR-PUFs, whose internal mathematical models are more complicated than other PUF constructions studied in this thesis. At the time of writing this dissertation, no such model has yet been known.

A BR-PUF consists of n stages (n is an even number), where each stage consists of two NOR gates, one demultiplexer and one multiplexer, see Fig. 6.1. Based on the value of the ith bit of a challenge applied to the ith stage, one of the NOR gates is selected. By setting the reset signal to low, the signal propagates in the ring. The final state of the inverter ring is a function of the gains and the propagation delays of the gates [32]. The response of the PUF is a binary value, which can be read from a predefined location on the ring between two stages, see Fig. 6.1. When applying a challenge, the ring may settle at a stable state after an oscillation period. However, for a specific set of challenges, the ring might stay in the metastable state for an infinite time, and the oscillation can be observed in the output of the PUF.

As suggested in [32], one might be able to extend the electrical model of the SRAM circuit in the metastable state and analyze the behavior of the inverter ring. However, it is a very challenging task due to the heavy dependence of such model on the physical characteristics of the primitive that embodies a BR-PUF (for more details see [32]).

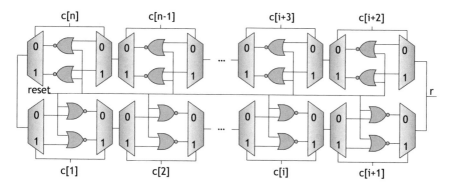

Fig. 6.1 The schematic of a BR-PUF with n stages. The response of the PUF can be read between two arbitrary stages. For a given challenge, the reset signal can be set low to activate the PUF. After a transient period , the BR-PUF might be settled to an allowed logical state

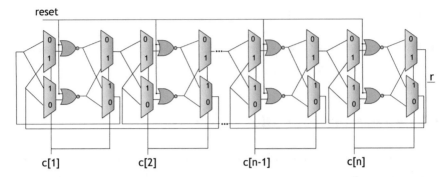

Fig. 6.2 The schematic of a TBR-PUF with n stages. The response of the PUF is read after the last stage. For a given challenge, the reset signal can be set low to activate the PUF. After a transient period, the BR-PUF might be settled to an allowed logical state

Although the mathematical model of the functionality of a BR-PUF is unknown, it has been observed that this construction is vulnerable to simple linear approximations [101]. Hence, the notion of TBR-PUF, as an enhancement to BR-PUFs, has been introduced [101]. Similar to BR-PUFs, a TBR-PUF consists of n stages (n is an even number), where each stage consists of two NOR gates. In contrast to BR-PUF, where for a given challenge only one of the NOR gates in each stage is selected, all $2n$ gates are chosen to contribute to the final response of a TBR-PUF. This can be achieved by placing two multiplexers before and two multiplexers after each stage and having feedback lines between different stages, see Fig. 6.2. As all the NOR gates are always in the circuit, the challenge-specific bias can be reduced.

6.3   A Constant Upper Bound on the Number of Influential Bits

This section provides experimental results supporting the following claim. The number of influential bits of the Boolean function representing a BR-PUF is upper bounded by a constant value.

6.3.1   Heuristic Approaches

First, we reflect the fact that our Theorem 2.3.2 is in line with the empirical results obtained by applying heuristic approaches, reported in [101, 121].

In an attempt to assess the security of BR-PUFs, Yamamoto et al. have implemented BR-PUFs on several FPGAs to analyze the influence of challenge bits on the respective responses [121]. They have explicitly underlined the existence of

Table 6.1 Statistical analysis of the 2048 CRPs, given to a 64-bit BR-PUF [121]. The first column shows the rule found in the samples, whereas the second column indicates the estimated probability of predicting the response

Rule	Est. Pr.
$(c_1 = 0) \rightarrow y = 1$	0.684
$(c_9 = 0) \wedge (c_6 = 1) \rightarrow y = 1$	0.762
$(c_{25} = 0) \wedge (c_{18} = 1) \wedge (c_1 = 0) \rightarrow y = 1$	0.852
$(c_{27} = 0) \wedge (c_{25} = 0) \wedge (c_{18} = 1) \wedge (c_6 = 1) \rightarrow y = 1$	0.932
$(c_{53} = 0) \wedge (c_{51} = 0) \wedge (c_{45} = 0) \wedge (c_{18} = 1) \wedge (c_7 = 0) \rightarrow y = 1$	1

influential bits and found so-called prediction rules. Table 6.1 summarizes their results, where for each type of the rules (monomials of different sizes) we report only the one with the highest estimated response prediction probability. In addition to providing evidence for the existence of influential bits, the size of the associated monomials is of particular importance for us. As shown in Table 6.1, their maximum size is surprisingly small, i.e., only five.

Similarly, the authors of [101] translate the influence of the challenge bits to the weights needed in artificial neural networks that represent the challenge-response behavior of BR-PUFs and the TBR-PUFs. They observed that there is a pattern in these weights, which models the influence of the challenge bits. It clearly reflects the fact that there exist influential bits determining the response of the PUF to a given challenge. From the results presented in [101], we conclude that there exists at least one influential bit. Nevertheless, the precise number of influential bits has not been further investigated in [101].

Inspired by the above results from [101, 121], we conduct further experiments. We collect 30000 CRPs from BR-PUFs and TBR-PUFs implemented on Altera Cyclone IV FPGAs (60 nm technology). In all of our PUF instances at least one influential bit is found, and the maximum number of influential bits (corresponding to the size of the monomials) is just a constant value in all cases. For the sake of readability, we present here only the results obtained for one arbitrary PUF instance. Our results shown in Table 6.2 are not only aligned with the results reported in [101, 121], but also reflect our previous theoretical findings. We could conclude this section as follows. At least one influential bit determines the response of a BR-PUF (respectively, TBR-PUF) to a given challenge. However, for the purpose of our framework their existence is not enough, and we need an upper bound on the number of influential bits.

Looking more carefully into the three different datasets, namely our own and the data reported in [101, 121], we observe that the total number of influential bits is always only a small value.

Table 6.2 Our statistical analysis of the 30000 CRPs, given to a 64-bit BR-PUF. The first column shows the rule found in the sample, whereas the second column indicates the estimated probability of predicting the response

Rule	Est. Pr.
$(c_{61} = 1) \rightarrow y = 1$	0.71
$(c_{11} = 0) \rightarrow y = 1$	0.72
$(c_{29} = 1) \rightarrow y = 1$	0.725
$(c_{39} = 0) \rightarrow y = 1$	0.736
$(c_{23} = 1) \rightarrow y = 1$	0.74
$(c_{50} = 1) \rightarrow y = 1$	0.745
$(c_{46} = 1) \rightarrow y = 1$	0.75
$(c_{61} = 1) \wedge (c_{23} = 1) \rightarrow y = 1$	0.82
$(c_{61} = 1) \wedge (c_{11} = 0) \rightarrow y = 1$	0.80
$(c_{23} = 1) \wedge (c_{46} = 1) \rightarrow y = 1$	0.86
$(c_{39} = 1) \wedge (c_{50} = 1) \rightarrow y = 1$	0.85
$(c_{61} = 1) \wedge (c_{11} = 0) \wedge (c_{29} = 1) \rightarrow y = 1$	0.88
$(c_{50} = 1) \wedge (c_{23} = 1) \wedge (c_{46} = 1) \rightarrow y = 1$	0.93
$(c_{50} = 1) \wedge (c_{23} = 1) \wedge (c_{46} = 1) \wedge (c_{39} = 0) \rightarrow y = 1$	0.97
$(c_{50} = 1) \wedge (c_{23} = 1) \wedge (c_{11} = 0) \wedge (c_{39} = 0) \wedge (c_{29} = 1) \rightarrow y = 1$	0.98
$(c_{50} = 1) \wedge (c_{23} = 1) \wedge (c_{46} = 1) \wedge (c_{39} = 0) \wedge (c_{29} = 1) \rightarrow y = 1$	0.99
$(c_{50} = 1) \wedge (c_{23} = 1) \wedge (c_{46} = 1) \wedge (c_{39} = 0) \wedge (c_{29} = 1) \wedge (c_{11} = 0) \rightarrow y = 1$	0.994
$(c_{50} = 1) \wedge (c_{23} = 1) \wedge (c_{46} = 1) \wedge (c_{39} = 0) \wedge (c_{29} = 1) \wedge (c_{61} = 0) \rightarrow y = 1$	0.995
$(c_{50} = 1) \wedge (c_{23} = 1) \wedge (c_{46} = 1) \wedge (c_{39} = 0) \wedge (c_{29} = 1) \wedge (c_{61} = 1) \wedge (c_{11} = 0) \rightarrow y = 1$	1

6.3.2   A Boolean-Analytical Approach

The commonly observed phenomenon reported in Sect. 6.3.1 should be further discussed from the point of view of Boolean analysis. This analysis enables us to study the properties of Boolean functions representing BR-PUFs more carefully, and consequently, to reflect what has been observed in practice. As discussed in Sect. 2.3.3, Kahn's theorem gives a hint that there exist at least one bit position with the influence $\Omega(p \log_2(n)/n)$. Although this theorem implies the existence of influential bits, it does quantify neither the number of influential bit positions nor the average sensitivity (see Sect. 2.3.2). Hence, for a family of PUFs, e.g., BR-PUFs, it is necessary to compute the average sensitivity of a Boolean function representing the respective PUFs. In line with this approach, at first stage, we compute for our PUFs (implemented on FPGAs) the average sensitivity of their respective Boolean functions.

As explained in Sect. 2.3.1, for a Boolean function f, the influence of a variable and the total average sensitivity can be calculated by employing Fourier analysis. However, in practice this analysis is computationally expensive. Instead, it suffices to simply approximate the respective average sensitivity. This idea has been extensively studied in the learning theory- and property testing-related literature (see [54], for a survey). Here we describe how the average sensitivity of a Boolean function, representing a PUF, can be approximated. We follow the simple and effective algorithm as explained in [91]. The central idea behind their algorithm is to collect a sufficient number of random pairs of labeled examples from the Boolean function, which are chosen uniformly at random. These pairs have the following property: $(c, f(c))$ and $(c^{\oplus i}, f(c^{\oplus i}))$, i.e., the inputs differ on a single Boolean variable. In other words, we first draw a pair $(c, f(c))$ uniformly at random, then by flipping the ith bit of the challenge and querying the function on this new challenge $c^{\oplus i}$ we collect two pairs $(c, f(c))$ and $(c^{\oplus i}, f(c^{\oplus i}))$. According to the definition of the influence of the ith variable on our Boolean function, $\text{Inf}_i(f)$, if we collect a large enough number of such pairs we can approximate the influence $\text{Inf}_i(f)$. Repeating this process for all the challenge bits, we can approximate the average sensitivity $I(f)$. Clearly, with probability $I(f)/n$ a pair chosen uniformly at random can be influential, i.e., $f(c) \neq f(c^{\oplus i})$. Therefore, it can easily be shown that in order to obtain an ε-approximator of $I(f)$ the number of queries to the function f is $O(p(1/\varepsilon)n/I(f))$ [91]. Note that the sample complexity of computing the average sensitivity is irrelevant to the complexity of our PAC learning framework (see Sect. 6.4).

To compute the average sensitivity of Boolean functions representing BR-PUFs, we collect CRPs from these PUFs implemented on Altera Cyclone IV FPGAs. The pairs of CRPs, namely $(c, f(c))$ and $(c^{\oplus i}, f(c^{\oplus i}))$, are collected as explained before and given to a simple algorithm computing the average sensitivity that is implemented in Matlab [77]. The number of CRPs collected from BR-PUFs has been reported in Table 6.3. For n-bit BR-PUFs, where $4 \leq n \leq 16$, the complete set of CRPs has been collected, whereas for 32-bit and 64-bit BR-PUFs only subsets of CRPs have been required. Moreover, averaging over many instances of our BR-PUFs, we obtain the results shown in Table 6.3 (TBR-PUFs scored similarly). This striking result leads us to the following plausible heuristic.

"Constant Average Sensitivity of BR-PUF family": *for all practical values of n it holds that the average sensitivity of a Boolean function associated with a physical n-bit PUF from the BR-PUF family is only a constant value.*

Although this heuristic has been partially proved, we provide evidence supporting our claim, namely, the experimental results and the results of Boolean analyses. Note

Table 6.3 The average sensitivity of n-bit BR-PUFs and the number of CRPs collected to compute the average sensitivity

n	# CRPs	The average sensitivity
4	16	1.25
8	256	1.86
16	65536	2.64
32	80000	3.6
64	160000	5.17

that the main cause of small, constant average sensitivity exhibited by the BR-PUF family has not yet been found; otherwise it could be possible to come up with a mathematical model representing BR-PUFs.

Finally, some relations between the average sensitivity and the strict avalanche criterion (SAC) can be recognized, although we believe that the average sensitivity is a more direct metric to evaluate the security of PUFs under ML attacks.

6.4 PAC Learning of PUFs Without Prior Knowledge of Their Mathematical Model

When discussing the PAC learnability of PUFs as target concepts, two scenarios should be distinguished. First, the precise mathematical model of the PUF functionality is known, and hence, a hypothesis representation is known to learn the PUF. This scenario has been considered in several studies, e.g., [34–36], where different hypothesis representations have been presented for each PUF family. Second, due to the lack of a precise mathematical model of the respective PUF functionality, to learn the PUF a more sophisticated approach is required. Therefore, the following question arises: is it possible to PAC learn a PUF family, even if we have no mathematical model of the physical functionality of the respective PUF family? We answer this question at least for the BR-PUF family. Figure 6.3 illustrates our roadmap for answering this question, more specifically, the steps that are taken to prove the PAC learnability of the BR-PUF family in the second scenario. While theoretical insights into the existence of influential bits, i.e., the first block, have been presented in Sect. 2.3.3, Sect. 6.3 provides more specific results for the BR-PUF family. According to these new insights, here we eventually prove that BR-PUF family, which lack a precise mathematical model, can nevertheless be PAC learned (see last two blocks in Fig. 6.3).

Weak Learning and Boosting of BR-PUFs

The key idea behind our learning framework is the provable existence of influential bits for PUFs and the constant average sensitivity of BR-PUFs in our scenario. These facts are taken into account to prove the existence of weak learners for the BR-PUF family. We start with Theorem 6.4.1 proved by Friedgut [30].

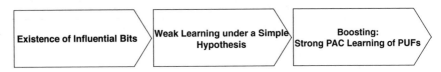

Fig. 6.3 Our roadmap for proving the PAC learnability of BR-PUF family, whose mathematical model is unknown

Theorem 6.4.1 *Every Boolean function $f : \{0, 1\}^n \rightarrow \{0, 1\}$ with $I(f) = k$ can be ε-approximated by another Boolean function h depending on only a constant number of Boolean variables K, where $K = \exp\left((2 + \sqrt{2\varepsilon}\log_2(4k/\varepsilon)/k)\frac{k}{\varepsilon}\right)$, and $\varepsilon > 0$ is an arbitrary constant.*

The implication of this result is that Boolean functions with low average sensitivity meet a sufficient condition for being close to a junta (see Definition 2.3.1). In addition to the impact of Theorem 6.4.1 on the machine learning theory, it has been well acknowledged in the domain of property testing algorithms. This field of study mainly concerns the problem of determining whether a function f is involved in a specific class of Boolean functions, for instance the class of K-junta functions (see Sect. 2.3 for the definition of K-juntas) [86]. More specifically, by querying the function $f : \{0, 1\}^n \rightarrow \{0, 1\}$, an algorithm can make a decision whether f is ε-close to a function $h : \{0, 1\}^n \rightarrow \{0, 1\}$ that is K-junta, i.e.,

$$\Pr[f(c) \neq h(c)] \leq \varepsilon.$$

Among all such algorithms (see [41] for a survey), an interesting algorithm proposed by Guijarro et al. bridges between two domains, namely PAC learning and property testing [43]. Their algorithm draws in an *adaptive* manner m examples of the function f as follows. First, an example is drawn randomly, afterwards this example is modified by flipping the variables one by one, and then querying the function f on the obtained examples. It has been proved that the number of queries to the function f is $O\left(K(1/\varepsilon\log(K/\delta) + \log n)\right)$ [43].

In order to examine whether the Boolean function f representing a BR-PUF belongs to the class of K-junta functions, we implement the algorithm proposed in [43] in Matlab [77]. As described above, the algorithm must have access to the examples, which are not collected in advance, but during the execution of the algorithm the function is queried frequently. Therefore, a Python script is developed that serves as an interface between the algorithm programmed in Matlab and the BR-PUF implemented on Altera Cyclone IV FPGAs. The FPGA communicates with the Python script through a Universal Asynchronous Receiver/Transmitter (UART). Matlab codes are then sent to the FPGA as commands, when executing our Python script. In our experiments we set $\delta = 0.001$ and $\varepsilon = 0.49$. While the former indicates that with probability $1 - \delta = 0.999$ the algorithm delivers a K-junta, the latter denotes the accuracy of the K-junta approximation the function f. Note that this setting is tailored to the specific needs of our PAC learner framework (see Sect. 6.4). Table 6.4 lists the results of our experiments for different BR-PUFs.

We explain now how Theorem 6.4.1 in conjunction with the results presented in Sect. 6.3 help us to prove the existence of a weak learner (Definition 2.6.3) for the BR-PUF family.

Theorem 6.4.2 *Every PUF from the BR-PUF family is weakly learnable.*

Proof For an arbitrary PUF from the BR-PUF family, consider its associated but unknown Boolean function that is denoted by f_{PUF} (i.e., our target concept). Our

Table 6.4 The results of our experiments examining whether a BR-PUF belongs to the class of K-junta functions

n	# CRPs	K
4	16	1
8	77	4
16	169	5
32	372	5
64	811	7

weak learning framework has two main steps. In the first step, we identify a weak approximator for f_{PUF}, which can only guarantee that the total error of the learner does not exceed $1/2 - \gamma$ ($\gamma > 0$). In the second phase this approximator is PAC learned (in a strong sense).

The first step relies on the fact that we can upper bound $I(f_{PUF})$ by some small constant value k due to the Constant Average Sensitivity heuristic. Now Theorem 6.4.1 ensures that there is a Boolean function h that is an ε-approximator of f_{PUF}, which depends only on a constant number of Boolean variables K since k and ε are constant values, independent of n. However, note that h depends on an unknown set of K variables. Thus, our Boolean function h is a so-called K-junta function, cf. [84]. More importantly, for constant K it is known that the K-junta function can be PAC learned by a trivial algorithm within $O\left(n^K\right)$ steps, cf. [5, 10, 13]. This PAC algorithm is indeed our algorithm WL that weakly learns f_{PUF}. Carefully choosing the parameters related to our approximators as well as the PAC learning algorithm, we ensure that WL returns a $1/2 - \gamma$-approximator for f_{PUF} and some $\gamma > 0$.   ∎

Applying now the canonical booster introduced in Sect. 2.6 to our WL proposed in the proof of Theorem 6.4.2, according to Definition 2.6.4, our weak learning algorithm can be transformed into an efficient and strong PAC learning algorithm.

Corollary 6.4.1 *BR-PUFs are strong PAC learnable, regardless of any mathematical model representing their challenge-response behavior.*

6.5 Results and Discussion

We have implemented BR and TBR-PUFs with 64 stages on an Altera Cyclone IV FPGA, manufactured on a 60 nm technology [3].[1]

As discussed and proved in Sect. 6.4, having influential bits enables us to define a prediction rule, which can serve as a hypothesis representation fulfilling the requirements of a weak learner. The algorithm WL proposed in the proof of the Theorem 6.4.2

---

[1]Further information on the implementation of these PUFs, the approach applied to deal with noisy responses and to identify PUFs with a minimum bias can be found in [32].

relies on the PAC learnability of K-juntas, where K is a small constant. However, it is known that every efficient algorithm for learning K-DTs (i.e., the number of leaves is 2^K) is an efficient algorithm for learning K-juntas, see, e.g., [80]. Furthermore, it is known that DLs generalize K-DTs [90]. Moreover, a monomial $M_{n,K}$ is a very simple type of a K-junta, where only the conjunction of the relevant variables is taken into account. Therefore, for our experiments we decide to let our weak learning algorithms deliver DLs, Monomials, and DTs.

To learn the challenge-response behavior of BR- and TBR-PUFs using these representations, we use the open source machine learning software Weka [44]. One may argue that more advanced tools might be available, but here we only aim to demonstrate that publicly accessible, and off-the-shelf software can be used to launch our proposed attacks. All experiments are conducted on a MacBook Pro with 2.6 GHz Intel Core i5 processor and 8GB of RAM. To boost the prediction accuracy of the model established by our weak learners, we apply the Adaptive Boosting (AdaBoost) algorithm [28]; nevertheless, any other boosting framework can be employed as well. For Adaboost, it is known that the error of the final model delivered by the boosted algorithm after T iteration is theoretically upper bounded by $\prod_{t=1}^{T} \sqrt{1 - 4\gamma^2}$, cf. [100]. To provide a better understanding of the relation between K, the number of iterations, and the theoretical bound on the error of the final model, a corresponding graph is shown in Fig. 6.4. We stress that this theoretical bound is solely asymptotically (in T) meaningful. More specifically, as can be seen from the graph, for very few iterations of the boosting algorithm the error of the final model delivered by the boosted algorithm is close to 1 (cf. [100, pp. 57–60], and Fig. 3.1 in that). However, it is known that we start from a weak learner, whose error rate is strictly below 0.5. This can be explained by an Adaboosts's theoretical worst-case analysis resulting in a rather loose, asymptotically correct bound [100].

Our experiments in Weka consist of a training phase and a testing phase. In the training phase, a model is established from the training data based on the chosen representation. Afterwards, the established model is evaluated on the test set, which contains an unseen subset of CRPs. Accordingly, the relevant question is about the

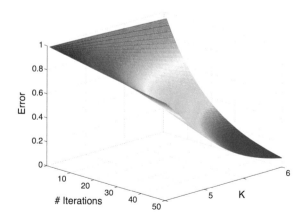

Fig. 6.4 The relation between the theoretical upper bound on the error of the final model returned by Adaboost, the number of iterations, and K. The graph is plotted for $k = 2$, $\varepsilon' = 0.01$, and $n = 64$. Here, $\varepsilon' = 0.01$ denotes the error of the K-junta learner

minimum number of CRPs required by the weak learner to deliver the model. As discussed in Sect. 2.6, this minimum number is determined by not only the accuracy and the confidence levels, but also the Vapnik-Chervonenkis dimension of our target concept class F that is the class of K-juntas. In [2] it has been proved that the lower bound on the number of examples needed by an algorithm to PAC learn a K-junta is

$$\Omega \left(1/\varepsilon \ln 1/\delta + K \ln 2 \ln n + 2^K \right).$$

Regarding the results of our experiments, presented in Sect. 6.3.1, if a single bit of the challenges determines the response of our 64-bit BR-PUF, the accuracy of the approximator is better than 50%. More precisely, the accuracy of a 1-junta ($K = 1$) as an ε-approximator of the Boolean function representing our BR-PUF is better than 50%. In other words, a 1-junta can be an appropriate representation, when weakly PAC learn our BR-PUF. Therefore, in order to compute the minimum number of examples needed by our weak learner, as an example we can set: $K = 1$, $\varepsilon = 0.49$, and $\delta = 0.01$. In this setting, the minimum number of examples is 15.

We evaluate this in practice by conducting experiments, where the size of the training set is $m = 15$, whereas the test set contains 30000 CRPs. The corresponding results presented in Table 6.5. As can be seen in this table, for all the representations (DLs, Monomials, and DTs), the accuracy of the models exceeds 50% as required for weak learners. Moreover, without boosting the accuracy of the models delivered by the weak learner is the same. This can be explained by the fact that giving solely 15 examples to the algorithm, the simple model delivered by it is a primitive rule that can be represented by each of the three representations.

For both BR-PUFs and TBR-PUFs, in each row of Table 6.5, the accuracy of DTs and DLs delivered by the algorithm is the same. This means that the rules found by the algorithm can be represented by DTs and DLs, where they share the same properties. By applying Adaboost the accuracy of the model is improved, however even after 50 iterations, the accuracy is not sufficiently high. This is due to the model delivered

Table 6.5 Experimental results for learning 64-bit BR-PUF and TBR-PUF, when $m = 15$. The accuracy $(1 - \varepsilon)$ is reported for three weak learners. The first row shows the accuracy of the weak learner, whereas the other rows show the accuracy of the boosted learner

# boosting iterations	BR-PUF			TBR-PUF		
	M_n (%)	DT (%)	DL (%)	M_n (%)	DT (%)	DL (%)
0 (No Boosting)	53.9	53.9	53.9	59.46	59.46	59.46
10	57.95	56.29	56.29	65.67	61.32	61.32
20	57.35	62.86	62.86	65.42	60.39	60.39
30	58.04	63.34	63.34	67.03	62.74	62.74
40	60.66	64.77	64.77	67.03	63.36	63.36
50	60.94	64.03	64.03	67.03	65.27	65.27

by the weak learner that is not expressive enough as the number of examples given to the algorithm is quite limited. Another interesting phenomenon that can be seen in a set of results listed in Table 6.5 is related to a fluctuation in the accuracy of the boosted model on its different iterations. This is a known phenomenon related to the accuracy of the model obtained after each iteration as well as how the models obtained in previous rounds are combined to obtain the final model, see Algorithm 1.1 (for more details see [100]).

We repeat our experiments for the case that the size of the training sets are $m = 100$ and $m = 1000$ (the size of the test set is 30000 CRPs). These experiments demonstrate that the weak learning of our test set always results in the delivery of a model with more than 50% accuracy as shown in the first rows of Tables 6.6 and 6.7. By boosting the respective models with AdaBoost, the accuracy is dramatically increased, see Tables 6.6 and 6.7. It can be observed that after 50 iterations of Adaboost applied to the weak model generated from 100 CRPs, the prediction accuracy of the boosted model is increased to more than 80% for all three representations. By increasing

Table 6.6 Experimental results for $m = 100$ (the same setting as for the Table 6.5)

# boosting iterations	BR-PUF			TBR-PUF		
	M_n (%)	DT (%)	DL (%)	M_n (%)	DT (%)	DL (%)
0 (No Boosting)	54.48	66.79	67.24	65.18	72.29	74.84
10	67.12	74.25	76.99	76.96	79.22	81.36
20	77.53	80.53	80.89	82.05	85.73	86.71
30	81.32	83.13	83.14	84.93	88.34	89.4
40	82.65	83.91	84.6	88.11	89.67	90.22
50	82.65	85.62	85.5	90.05	89.69	91.58

Table 6.7 Experimental results for $m = 1000$ (the same setting as for the Tables 6.5 and 6.6)

# boosting iterations	BR-PUF			TBR-PUF		
	M_n (%)	DT (%)	DL (%)	M_n (%)	DT (%)	DL (%)
0 (No Boosting)	63.73	75.69	84.59	64.9	75.6	84.34
10	81.09	85.49	94.2	79.9	87.12	95.05
20	89.12	91.08	96.64	88.28	91.57	97.89
30	93.24	93.24	97.50	93.15	93.9	98.75
40	95.69	94.28	97.99	96.73	95.05	99.13
50	96.80	95.04	98.32	98.4	95.96	99.37

the number of samples to 1000 CRPs, the prediction accuracy is further increased up to 98.32% for learning the BR-PUFs, and 99.37% for learning the TBR-PUFs under DL representations. It is interesting to observe that the simplest representation class, i.e., Monomials clearly presents the greatest advantage given by the boosting technique. As explained in [100] this is due to avoiding any overfitting tendency.

Chapter 7
Follow-Up Work

> *[...] our [hybrid] attacks have only cubic complexity. This is a drastic improvement over the exponential complexity of state-of-the-art, pure modeling attacks.*
>
> [97]

This section gives a brief introduction to our work derived from and built upon the results of the PAC learning framework established in this thesis. The complete description of these work and their results have been first published in [31, 110]. Here we merely focus on the key goals of these studies and methodologies applied to achieve them.

Our studies presented in [31, 110] share the following common features. First, they investigate the feasibility of launching a so-called *hybrid* attack against XOR Arbiter PUFs, where we combine the photonic emission analysis with either machine learning techniques or a lattice basis reduction attack. Although being costly, photonic emission analysis is a promising solution leading to a complete and linear characterization of XOR Arbiter PUFs, as shown by [109]. Second, in both attack scenarios, no limit on the number of arbiter chains has been assumed, and hence, the proposed attacks can be launched against XOR Arbiter PUFs with a large number of arbiter chains (e.g., more than 16). Following sections explain the principle behind these attacks in brief.

7.1 Lattice Basis Reduction Attack

As discussed in Chap. 4, XOR Arbiter PUFs are subject to ML attacks [35, 95]. The number of arbiter chains can be a limiting factor for this type of attacks, hence, the attack is effective only against XOR Arbiter PUFs with a small number of arbiter chains. Moreover, it has also been shown that when the number of arbiter chains exceeds this limited number, additional side channel analyses can be combined with

© Springer International Publishing AG 2018

F. Ganji, *On the Learnability of Physically Unclonable Functions*, T-Labs
Series in Telecommunication Services, https://doi.org/10.1007/978-3-319-76717-8_7

an ML attack to compromise the security of an XOR Arbiter PUF [97]. Otherwise, as proved in Chap. 4 and previously observed in practice (e.g., see [114]), the number of CRPs, and consequently the time, required to launch an ML attack increases drastically.

On the other hand, it has been demonstrated that Arbiter as well as XOR Arbiter PUFs can be physically characterized by performing a high resolution temporal photonic emission analysis [46, 109]. This attack does not require the response of the PUF, and the number of required measurements is only linear in the number of stages. Nevertheless, the attacker needs to apply the desired challenge. Similar to ML attacks, it is clear that when the attacker cannot rely on the direct control of the challenges (e.g., controlled PUF [37], Slender PUF [76], and Recyclable PUFs [55]), this attack would obviously fail.

On the basis of the above facts and considerations, the work presented in [31] aims to prove that an increase in the number of arbiter chains is not a *silver bullet*, but rather a very limited step towards improving the robustness of XOR Arbiter PUFs against attacks. In this regard, we present a novel mathematical proof of the vulnerability of legacy and controlled XOR Arbiter PUFs to lattice basis reduction attacks. Specifically, we demonstrate how the lattice basis reduction attack proposed by Nguyen et al. [82] can be extended and combined with a photonic side channel analysis to compromise the security of XOR Arbiter PUFs.

Thanks to the method developed in a series of work by Tajik et al. (see, e.g., [108, 109]), our methodology of collecting physically measured data from an XOR Arbiter PUF is nowadays very well accepted. This data is fed into the algorithm that runs lattice basis reduction attack against the corresponding XOR Arbiter PUF, represented by a hidden sub set sum problem [16]. As a result, our attack can disclose the *hidden* challenges applied to a controlled XOR Arbiter PUF, composed of an arbitrary, large number of arbiter chains, and thus break the security of this PUF family without knowing the challenges or their corresponding responses.

In a nutshell, we demonstrate that neither an increase in the number of arbiter chains nor the limitation of the access to challenges and responses qualifies as a secure countermeasure against our hybrid attack.

7.2 Laser Fault Injection Attack

One of the main goals of the study presented in [110] is to simplify a modeling attack against XOR Arbiter PUFs featuring a large number of arbiter chains. To achieve this goal, our study discusses how an attack can take advantage of results achieved by mounting semi-invasive attacks, e.g., [109, 111]. In this study, we rely on the fact that different hardware primitives, specially PUF circuits, can be identified and localized on programmable logic devices (PLDs). After localizing a PUF circuit, it is shown how the building blocks of the PUF can be manipulated by inducing faults into the logic cells to change the configuration of the PUF.

As shown in Chap. 4, the learning complexity of an XOR Arbiter PUF is exponential in the number of arbiter chains. We demonstrate how deactivating multiple arbiter chains in an XOR Arbiter PUF enables the attacker to come up with a model of the PUF in a polynomial time, regardless of the number of chains. To this end, we propose to induce a fault into all the arbiter chains, except an arbiter chain which has to be learned. More precisely, when having access to the response of an arbiter chain of the XOR Arbiter PUF, that chain can be PAC learned separately in polynomial time, e.g., following the procedure in Chap. 3. By obtaining a model of the challenge-response behavior of each and every chain individually, the response of the XOR Arbiter PUF to an unseen challenge can be predicted, i.e., the security of the XOR Arbiter PUF can be compromised. Accordingly, the maximum number of CRPs required by our hybrid attack is polynomial in the number of arbiter chains, which is significantly improved in comparison to the pure PAC learning attack.

The main conclusion that can be made from the results of the studies introduced in this chapter is that countermeasures against ML attacks cannot ensure the security of PUFs against hybrid attacks. To launch such a hybrid attack, we make a set of assumptions regarding the access of an attacker to equipment employed to launch semi-invasive attacks. One can argue that such devices and equipment are not easily accessible, and therefore, launching such attacks seems infeasible in practice. We should stress that the equipment used in our studies can be rented, or even assembled by the attacker, as demonstrated in [108].

Another important aspect of the studies presented in [31, 110] is the practical application of PAC learning. Although being originally pure theoretical, PAC learning can be combined with an experimental analysis, i.e., a photonic emission analysis in our case, to improve the efficiency and effectiveness of the PAC learning framework.

Chapter 8
Conclusion and Future Work

> *Know your enemy and know yourself, find naught in fear for*
> *hundred battles. Know yourself, but not your enemy, find level of*
> *loss and victory. Know thy enemy, but not yourself, wallow in*
> *defeat every time.*
>
> The Art of War, Sun Tzu

Nowadays, PUFs are among the main building blocks of the security measures, which are considered necessary to not only mitigate attacks against ICs but also remedy the shortcomings of the corresponding traditional countermeasures. However, after more than a decade of the invention of PUFs, the design of a PUF fulfilling the necessary, and natural requirements is still a challenging task. Being unpredictable and unclonable are involved in these requirements, albeit being unsatisfied as demonstrated by the results of various attacks. ML attacks provide evidence supporting that an adversary can launch a cost-efficient attack to predict the response of the PUF to an unseen, arbitrarily chosen challenge. Although several studies focus on the security assessment of PUFs by mounting ML attacks, to the best of our knowledge, our study is the first work that addresses the issue with the final model delivery after the learning phase. In this regard, this thesis proposes a generic PAC learning framework, which has been successfully applied to known, and widely accepted families of PUFs, namely, Arbiter, XOR Arbiter, RO-, and BR-PUFs. More specifically, our framework clearly states how these PUF families can be learned, for given levels of accuracy and confidence. Interestingly enough, the maximum number of CRPs required to mount our attack can be calculated beforehand.

Two main steps have been taken within our framework. First, by exploring the inherent physical characteristics of the PUFs mentioned above, fit-for-purpose mathematical representations of them are established. We demonstrate that in order to adequately reflect the physical behavior of these primitives, it is crucial to explore the inherent physical properties of the PUFs and *translate* them into mathematical representations. To this end, we employ methods and techniques that have been known and widely accepted within ML community. Second, to learn the PUFs under the appropriate representations, polynomial time ML algorithms have been introduced.

© Springer International Publishing AG 2018

F. Ganji, *On the Learnability of Physically Unclonable Functions*, T-Labs
Series in Telecommunication Services, https://doi.org/10.1007/978-3-319-76717-8_8

In addition to the ML algorithms used to learn the challenge-response behavior of the PUFs, we apply a set of algorithms that have been originally adopted as property testing algorithms. Such algorithms can be employed to evaluate the security of PUFs and serve as a toolbox helping PUF designers and manufacturers with the identification of best practices.

Clearly, our ultimate goal is to provide a theoretical proof of the facts, which have been verified only empirically so far in the literature. Therefore, occasionally, no experimental proof-of-concept for our attack against some PUF families is presented. However, in such cases, for the sake of completeness, we have attempted to compare our theoretical finding with the results of experiments performed previously in the literature. We have shown that these results match the results of our rigorous mathematical approaches. Furthermore, in some attack scenarios, the feasibility and effectiveness of our machine learning attacks have also been evaluated by conducting extensive experiments on the PUFs implemented on real-world hardware platforms, similar to ones used in the most relevant literature.

Finally, we would like to stress that, when interpreting and applying the results of this study, the following issues should be taken into account.

The difference between PAC learning and empirical ML approaches. The key premise underlying a PAC learning framework is the guarantee of the final model delivery for prescribed levels of accuracy and confidence. In other words, when applying an empirical ML approach, none of the accuracy and the confidence levels can be defined beforehand. Moreover, the number of examples required to run an empirical algorithm is not known before the learning phase. These are in contrast to a PAC learning algorithm, where not only the accuracy and the confidence levels can be defined before running the ML algorithms, but also the number of examples required to achieve these levels can be computed.

Infeasibility of PAC learning and its consequences. Infeasibility of applying a PAC learning framework may refer to the lack of a polynomial-sized hypothesis representing a function, and/or existence of no polynomial time algorithm to learn a target concept or its associated representation. In both cases, empirical algorithms can be run to learn the target concept. Nevertheless, the accuracy and the confidence levels cannot be defined prior to launching the attack, and the number of CRPs required to be collected should be estimated in a trial and error process.

Generalization of the results of this study to other PUFs. Although it is tempting to generalize the results achieved in this thesis, these results should be interpreted carefully. In particular, the results of PAC learning approach applied to learn BR-PUF family cannot be generalized to other PUF families, even though it needs no mathematical model. As discussed in Chap. 6, the essence of PAC learnability of the BR-PUF family is the constant average sensitivity of these PUFs. To be precise, the feasibility and effectiveness of our attack depend heavily on the number of Boolean variables (K), which have an influence on the response of the PUF. More specifically, the limit on the feasibility of our attack has been determined by the algorithms that have been already developed in the literature to (weakly) PAC learn a K-junta. From a learning theory perspective, the problem of learning a K-junta in polynomial time is an open problem. Nonetheless, for constant K it is known that the K-junta function can be PAC learned by a trivial algorithm within $O\left(n^K\right)$ steps, cf. [5, 10, 13]. Hence, when K is not constant, the time complexity of the learning algorithm imposes a limit on the efficiency of our attack.

Last but not least, although our PAC learning framework has its own novelty value, we feel that the precise, mathematical description of the characteristics of PUFs studied in this thesis is the most important aspects of our work. We believe that this description can help to fill the gap between the mathematical design of cryptographic primitives and the design of PUFs in the real world. As evidence thereof, the Siegenthaler Theorem and the Fourier analysis that are well-known and widely used in modern cryptography may provide special insights into the physical design of secure PUFs in the future. Likewise, this thesis paves the way towards developing a more systematic approach to assessing the security of PUFs through providing proofs of learnability, originated in ML theory.

Future Research: The following outlines the possible, further research directions in the area of security assessment of PUFs.

- Another important aspect of our PAC learning framework that should be considered is how it deals with noisy CRPs. This problem has been partially addressed in Chap. 4, namely only for XOR Arbiter PUFs. In spite of that, this problem has been well studied in the ML community. More specifically, PAC learning and boosting algorithms can tolerate noise (see, e.g., [17, 53]). In the future, we will indeed investigate this important issue.
- As a result of attacks against PUFs, a paradigm shift from standalone PUFs (i.e., PUFs being not composed of a combination of some PUFs) to more sophisticated, composite PUF schemes has been aimed. As a prime example, Arbiter PUF manufacturers have been forced to put effort into the design and development of countermeasures including XOR Arbiter PUFs as an effective technical action. As discussed in Chap. 4, the security of XOR Arbiter PUFs with a limited number of arbiter chains can be broken by launching ML attacks. On the other hand, XOR Arbiter PUFs composed of a large number of arbiter chains are susceptible to hybrid attacks (i.e., the combination of ML and semi-invasive attacks). Therefore, it seems crucial and interesting to come up with new PUF schemes, composed of PUFs that are combined by mean of other combinatorial functions instead of XOR function, and their resilience against ML attacks can be proved.
- To evaluate the security of PUFs against ML attacks, we have introduced new notions and metrics, first introduced in Boolean analysis and property testing domains of study. Moreover, Fourier analysis has been conducted to reveal some *hidden* properties of the Boolean functions representing PUFs. These can contribute to the development of a new benchmarking system, which can include not only the previous metrics (e.g., robustness, uniqueness, etc.) but also new metrics originated in the Boolean analysis, for instance, the average sensitivity.

References

1. AG, I.T.: Spotlight on embedded security keeping up with the internet of things. http://www.infineon.com/dgdl/Infineon-Spolight_on_embedded_security-ABR-v10_15-EN.pdf?fileId=5546d462503812bb0150a397d7663251(2015). Accessed 22 Feb 2017
2. Almuallim, H., Dietterich, T.G.: Learning with many irrelevant features. In: Proceedings of the Ninth National Conference on Artificial Intelligence (1991)
3. Altera: Cyclone IV Device Handbook. Altera Corporation, San Jose (2014)
4. Angluin, D.: Learning regular sets from queries and counterexamples. Inf. Comput. **75**(2), 87–106 (1987)
5. Angluin, D.: Queries and concept learning. Mach. Learn. **2**(4), 319–342 (1988)
6. Angluin, D., Laird, P.: Learning from noisy examples. Mach. Learn. **2**(4), 343–370 (1988)
7. Anthony, M.: Computational Learning Theory. Cambridge University Press, Cambridge (1997)
8. Armknecht, F., Maes, R., Sadeghi, A., Standaert, O.X., Wachsmann, C.: A formalization of the security features of physical functions. In: IEEE Symposium on Security and Privacy (SP), pp. 397–412 (2011)
9. Armknecht, F., Moriyama, D., Sadeghi, A.R., Yung, M.: Towards a unified security model for physically unclonable functions. In: Topics in Cryptology-CT-RSA 2016: The Cryptographers' Track at the RSA Conference, vol. 9610, p. 271. Springer, Berlin (2016)
10. Arvind, V., Köbler, J., Lindner, W.: Parameterized learnability of k-juntas and related problems. In: Algorithmic Learning Theory, pp. 120–134. Springer, Berlin (2007)
11. Becker, G.T.: The gap between promise and reality: on the insecurity of XOR arbiter PUFs. In: International Workshop on Cryptographic Hardware and Embedded Systems. pp. 535–555. Springer, Berlin (2015)
12. Blum, A., Frieze, A., Kannan, R., Vempala, S.: A polynomial-time algorithm for learning noisy linear threshold functions. In: Proceedings of 37th Annual Symposium on Foundations of Computer Science. pp. 330–338. IEEE, New York (1996)
13. Blum, A.L., Langley, P.: Selection of relevant features and examples in machine learning. Artif. Intell. **97**(1), 245–271 (1997)
14. Blumer, A., Ehrenfeucht, A., Haussler, D., Warmuth, M.: Classifying learnable geometric concepts with the Vapnik-Chervonenkis dimension. In: Proceedings of the Eighteenth Annual ACM Symposium on Theory of Computing. pp. 273–282. ACM (1986)
15. Blumer, A., Ehrenfeucht, A., Haussler, D., Warmuth, M.K.: Learnability and the Vapnik–Chervonenkis dimension. J. ACM **36**(4), 929–965 (1989)
16. Boyko, V., Peinado, M., Venkatesan, R.: Speeding up Discrete Log and Factoring Based Schemes via Precomputations. pp. 221–235. Springer, Berlin (1998)

© Springer International Publishing AG 2018
F. Ganji, *On the Learnability of Physically Unclonable Functions*, T-Labs Series in Telecommunication Services, https://doi.org/10.1007/978-3-319-76717-8

17. Bshouty, N.H., Jackson, J.C., Tamon, C.: Uniform-distribution attribute noise learnability. Inf. Comput. **187**(2), 277–290 (2003)
18. Bylander, T.: Learning linear threshold functions in the presence of classification noise. In: Proceedings of the Seventh Annual Conference on Computational Learning Theory. pp. 340–347 (1994)
19. Chen, L., Franklin, J., Regenscheid, A.: Guidelines on hardware-rooted security in mobile devices (Draft). NIST Spec. Publ. **800**(164), 10–11 (2012)
20. Chen, Q., Csaba, G., Lugli, P., Schlichtmann, U., Rührmair, U.: The bistable ring PUF: a new architecture for strong physical unclonable functions. In: 2011 IEEE International Symposium on Hardware-Oriented Security and Trust (HOST), pp. 134–141. IEEE, New York (2011)
21. Coombs, R.: Securing the Future of Authentication with ARM TrustZone-based Trusted Execution Environment and Fast Identity Online (FIDO). ARM White Paper (2015)
22. Delvaux, J., Gu, D., Schellekens, D., Verbauwhede, I.: Secure lightweight entity authentication with strong PUFs: mission impossible? In: Cryptographic Hardware and Embedded Systems–CHES 2014, pp. 451–475. Springer, Berlin (2014)
23. Delvaux, J., Verbauwhede, I.: Fault Injection Modeling Attacks on 65nm Arbiter and RO Sum PUFs via Environmental Changes. Technical reports, Cryptology ePrint Archive: Report 2013/619. https://eprint.iacr.org/2013/619 (2013)
24. Delvaux, J., Verbauwhede, I.: Side channel modeling attacks on 65 nm arbiter PUFs exploiting CMOS device noise. In: IEEE International Symposium on Hardware-Oriented Security and Trust (HOST), pp. 137–142 (2013)
25. Ehrenfeucht, A., Haussler, D., Kearns, M., Valiant, L.: A general lower bound on the number of examples needed for learning. Inf. Comput. **82**(3), 247–261 (1989)
26. Fischer, P., Simon, H.U.: On learning ring-sum-expansions. SIAM J. Comput. **21**(1), 181–192 (1992)
27. Freund, Y.: Boosting a weak learning algorithm by majority. Inf. Comput. **121**(2), 256–285 (1995)
28. Freund, Y., Schapire, R.E.: A decision-theoretic generalization of on-line learning and an application to boosting. J. Comput. Syst. Sci. **55**(1), 119–139 (1997)
29. Freund, Y., Schapire, R.E.: Large margin classification using the perceptron algorithm. Mach. Learn. **37**(3), 277–296 (1999)
30. Friedgut, E.: Boolean functions with low average sensitivity depend on few coordinates. Combinatorica **18**(1), 27–35 (1998)
31. Ganji, F., Krämer, J., Seifert, J.P., Tajik, S.: Lattice basis reduction attack against physically unclonable functions. In: Proceedings of the 22nd ACM Conference on Computer and Communications Security. ACM (2015)
32. Ganji, F., Tajik, S., Fäßler, F., Seifert, J.P.: Strong machine learning attack against PUFs with no mathematical model. In: International Conference on Cryptographic Hardware and Embedded Systems–CHES. pp. 391–411. Springer, Berlin (2016)
33. Ganji, F., Tajik, S., Fäßler, F., Seifert, J.P.: Having no mathematical model may not secure PUFs. J. Cryptogr, Eng (2017)
34. Ganji, F., Tajik, S., Seifert, J.P.: Let me prove it to you: RO PUFs are provably learnable. In: The 18th Annual International Conference on Information Security and Cryptology (2015)
35. Ganji, F., Tajik, S., Seifert, J.P.: Why attackers win: on the learnability of XOR arbiter PUFs. In: Trust and Trustworthy Computing, pp. 22–39. Springer, Berlin (2015)
36. Ganji, F., Tajik, S., Seifert, J.P.: PAC learning of arbiter PUFs. J. Cryptogr. Eng. Spec. Sect. Proofs **2014**, 1–10 (2016)
37. Gassend, B., Clarke, D., Van Dijk, M., Devadas, S.: Controlled physical random functions. In: Proceedings 18th Annual Computer Security Applications Conference. pp. 149–160 (2002)
38. Gassend, B., Clarke, D., Van Dijk, M., Devadas, S.: Silicon physical random functions. In: Proceedings of the 9th ACM Conference on Computer and Communications Security. pp. 148–160 (2002)
39. Gassend, B., Lim, D., Clarke, D., Van Dijk, M., Devadas, S.: Identification and authentication of integrated circuits. Concurr. Comput. Pract. Exp. **16**(11), 1077–1098 (2004)

40. Gierlichs, B., Poschmann, A.Y.: Introduction to the CHES 2016 special issue. J. Cryptogr. Eng. pp. 1–2 (2017)
41. Goldreich, O.: Property Testing: Current Research and Surveys, vol. 6390. Springer, Berlin (2010)
42. Guajardo, J., Kumar, S.S., Schrijen, G.J., Tuyls, P.: FPGA intrinsic PUFs and their Use for IP protection. In: Cryptographic Hardware and Embedded Systems–CHES 2007, pp. 63–80. Springer, Berlin (2007)
43. Guijarro, D., Tarui, J., Tsukiji, T.: Finding relevant variables in PAC model with membership queries. In: International Conference on Algorithmic Learning Theory. pp. 313–322. Springer, Berlin (1999)
44. Hall, M., Frank, E., Holmes, G., Pfahringer, B., Reutemann, P., Witten, I.H.: The WEKA data mining software: an update. ACM SIGKDD Explor. Newsl. **11**(1), 10–18 (2009)
45. Hammouri, G., Öztürk, E., Sunar, B.: A tamper-proof and lightweight authentication scheme. Pervasive Mob. Comput. **4**(6), 807–818 (2008)
46. Helfmeier, C., Boit, C., Nedospasov, D., Seifert, J.P.: Cloning physically unclonable functions. In: IEEE International Symposium on Hardware-Oriented Security and Trust (HOST), pp. 1–6 (2013)
47. Helfmeier, C., Nedospasov, D., Tarnovsky, C., Krissler, J.S., Boit, C., Seifert, J.P.: Breaking and entering through the silicon. In: Proceedings of the 2013 ACM SIGSAC Conference on Computer & Communications Security. pp. 733–744. ACM (2013)
48. Helmbold, D., Sloan, R., Warmuth, M.K.: Learning integer lattices. SIAM J. Comput. **21**(2), 240–266 (1992)
49. Herder, C., Ren, L., Van Dijk, M., Yu, M.D., Devadas, S.: Trapdoor computational fuzzy extractors and stateless cryptographically-secure physical unclonable functions. In: IEEE Transactions on Dependable and Secure Computing (2016)
50. Herder, C., Yu, M.D., Koushanfar, F., Devadas, S.: Physical unclonable functions and applications: a tutorial. Proc. IEEE **102**(8), 1126–1141 (2014)
51. Hopcroft, J.E., Motwani, R., Ullman, J.D.: Automata Theory, Languages, and Computation. International Edition 24 (2006)
52. Kahn, J., Kalai, G., Linial, N.: The influence of variables on boolean functions. In: 29th Annual Symposium on Foundations of Computer Science, pp. 68–80. IEEE, New York (1988)
53. Kalai, A., Servedio, R.A.: Boosting in the presence of noise. In: Proceedings of the thirty-fifth annual ACM Symposium on Theory of Computing, pp. 195–205. ACM (2003)
54. Kalai, G., Safra, S.: Threshold phenomena and influence: perspectives from mathematics, computer science, and economics. In: Computational Complexity and Statistical Physics, St. Fe Institute, Studies in the science of complexity pp. 25–60 (2006)
55. Katzenbeisser, S., Kocabaş, Ü., Van Der Leest, V., Sadeghi, A.R., Schrijen, G.J., Wachsmann, C.: Recyclable PUFs: logically reconfigurable PUFs. J. Cryptogr. Eng. **1**(3), 177–186 (2011)
56. Kearns, M.J., Vazirani, U.V.: An Introduction to Computational Learning Theory. MIT press, USA (1994)
57. Khardon, R., Wachman, G.: Noise tolerant variants of the perceptron algorithm. J. Mach. Learn. Res. **8**, 227–248 (2007)
58. Kömmerling, O., Kuhn, M.: Design principles for tamper-resistant security processors. In: USENIX Workshop on Smartcard Technology (1999)
59. Koushanfar, F.: Hardware metering: a survey. In: Introduction to Hardware Security and Trust, pp. 103–122. Springer, Berlin (2012)
60. Koushanfar, F., Karri, R.: Can the SHIELD protect our integrated circuits?. In: 2014 IEEE 57th International Midwest Symposium on Circuits and Systems (MWSCAS), pp. 350–353 (2014)
61. Lee, J.W., Lim, D., Gassend, B., Suh, G.E., Van Dijk, M., Devadas, S.: A technique to build a secret key in integrated circuits for identification and authentication applications. Digest of Technical Papers. In: 2004 Symposium on VLSI Circuits, pp. 176–179 (2004)
62. Lim, D.: Extracting secret keys from integrated circuits, Master thesis, Massachusetts Institute of Technology (2004)

63. Lim, D., Lee, J.W., Gassend, B., Suh, G.E., Van Dijk, M., Devadas, S.: Extracting secret keys from integrated circuits. IEEE Trans. Very Large Scale Integr. (VLSI) Syst. **13**(10), 1200–1205 (2005)
64. Linial, N., Mansour, Y., Rivest, R.L.: Results on learnability and the Vapnik-Chervonenkis dimension. Inf. Comput. **90**(1), 33–49 (1991)
65. Littlestone, N.: Learning quickly when irrelevant attributes abound: a new linear-threshold algorithm. Mach. Learn. **2**(4), 285–318 (1988)
66. Littlestone, N.: From on-line to batch learning. In: Proceedings of the Second Annual Workshop On Computational Learning Theory, pp. 269–284 (1989)
67. Maes, R.: Physically Unclonable Functions: Constructions. Properties and Applications. Springer, Berlin (2013)
68. Mahmoud, A., Rührmair, U., Majzoobi, M., Koushanfar, F.: Combined Modeling and Side Channel Attacks on Strong PUFs. Technical reports, Cryptology ePrint Archive: Report 2013/632. https://eprint.iacr.org/2013/632 (2013)
69. Maiti, A., Casarona, J., McHale, L., Schaumont, P.: A large scale characterization of RO-PUF. In: 2010 IEEE International Symposium on Hardware-Oriented Security and Trust (HOST), pp. 94–99 (2010)
70. Maiti, A., Kim, I., Schaumont, P.: A robust physical unclonable function with enhanced challenge-response set. IEEE Trans. Inf. Forensics Secur. **7**(1), 333–345 (2012)
71. Maiti, A., Schaumont, P.: Improving the quality of a physical unclonable function using configurable ring oscillators. In: 2009 International Conference on Field Programmable Logic and Applications, pp. 703–707. IEEE, New York (2009)
72. Majzoobi, M., Dyer, E., Elnably, A., Koushanfar, F.: Rapid FPGA delay characterization using clock synthesis and sparse sampling. In: 2010 IEEE International Test Conference (ITC), pp. 1–10 (2010)
73. Majzoobi, M., Koushanfar, F., Devadas, S.: FPGA PUF using programmable delay lines. In: 2010 IEEE International Workshop on Information Forensics and Security (WIFS), pp. 1–6 (2010)
74. Majzoobi, M., Koushanfar, F., Potkonjak, M.: Lightweight secure PUFs. In: Proceedings of the 2008 IEEE/ACM International Conference on Computer-Aided Design, pp. 670–673 (2008)
75. Majzoobi, M., Koushanfar, F., Potkonjak, M.: Techniques for design and implementation of secure reconfigurable PUFs. ACM Trans. Reconfigurable Tech. Syst. (TRETS) **2** (2009)
76. Majzoobi, M., Rostami, M., Koushanfar, F., Wallach, D.S., Devadas, S.: Slender PUF protocol: a lightweight, robust, and secure authentication by substring matching. In: IEEE Symposium on Security and Privacy Workshops (SPW), pp. 33–44 (2012)
77. MATLAB: Version 9.0.0.341360 (R2016a). The MathWorks Inc., Natick, Massachusetts (2016)
78. Merli, D., Schuster, D., Stumpf, F., Sigl, G.: Side-channel Analysis of PUFs and Fuzzy extractors. In: Trust and Trustworthy Computing, pp. 33–47. Springer, Berlin (2011)
79. Mitchell, T.M.: Machine Learning (McGraw-Hill Intl Editions Computer Science Series). McGraw-Hill (1997)
80. Mossel, E., O'Donnell, R., Servedio, R.A.: Learning functions of k relevant variables. J. Comput. Syst. Sci. **69**(3), 421–434 (2004)
81. Mugali, K.C., Patil, M.M.: A novel technique of configurable ring oscillator for physical unclonable functions. Int. J. Comput. Eng. Appl. **9**(1) (2015)
82. Nguyen, P., Stern, J.: The hardness of the hidden subset sum problem and its cryptographic implications. In: Advances in Cryptology–CRYPTO'99. pp. 31–46. Springer, Berlin (1999)
83. Nguyen, P.H., Sahoo, D.P., Chakraborty, R.S., Mukhopadhyay, D.: Efficient attacks on robust ring oscillator PUF with enhanced challenge-response set. In: Proceedings of the 2015 Design, Automation & Test in Europe Conference & Exhibition, pp. 641–646. EDA Consortium (2015)
84. O'Donnell, R.: Analysis of Boolean Functions. Cambridge University Press, Cambridge (2014)
85. Papoulis, A., Pillai, S.U.: Probability, Random Variables, and Stochastic Processes. Tata McGraw-Hill Education (2002)

86. Parnas, M., Ron, D., Samorodnitsky, A.: Proclaiming dictators and juntas or testing Boolean formulae. In: Approximation. Randomization, and Combinatorial Optimization: Algorithms and Techniques, pp. 273–285. Springer, Berlin (2001)
87. Parno, B., McCune, J.M., Perrig, A.: Bootstrapping Trust in Modern Computers. Springer Science & Business Media (2011)
88. Parusiński, M., Shariati, S., Kamel, D., Xavier-Standaert, F.: Strong PUFs and their (Physical) unpredictability: a case study with power PUFs. In: Proceedings of the Workshop on Embedded Systems Security, p. 5 (2013)
89. Rioul, O., Solé, P., Guilley, S., Danger, J.L.: On the entropy of physically unclonable functions. In: 2016 IEEE International Symposium on Information Theory (ISIT), pp. 2928–2932 (2016)
90. Rivest, R.L.: Learning decision lists. Mach. Learn. 2(3), 229–246 (1987)
91. Ron, D., Rubinfeld, R., Safra, M., Samorodnitsky, A., Weinstein, O.: Approximating the influence of monotone Boolean functions in $O(\sqrt{n})$ query complexity. ACM Trans. Comput. Theory (TOCT) 4(4), 11 (2012)
92. Rostami, M., Koushanfar, F., Karri, R.: A primer on hardware security: models, methods, and metrics. Proc. IEEE 102(8), 1283–1295 (2014)
93. Rostami, M., Majzoobi, M., Koushanfar, F., Wallach, D., Devadas, S.: Robust and reverse-engineering resilient PUF authentication and key-exchange by substring matching. IEEE Trans. Emerg. Top. Comput. 2(1), 37–49 (2014)
94. Rührmair, U., Busch, H., Katzenbeisser, S.: Strong PUFs: models, constructions, and security proofs. In: Towards Hardware-intrinsic Security, pp. 79–96. Springer, Berlin (2010)
95. Rührmair, U., Sehnke, F., Sölter, J., Dror, G., Devadas, S., Schmidhuber, J.: Modeling attacks on physical unclonable functions. In: Proceedings of the 17th ACM Conference on Computer and Communications Security, pp. 237–249 (2010)
96. Rührmair, U., Sölter, J., Sehnke, F.: On the Foundations of Physical Unclonable Functions. IACR Cryptology ePrint Archive 2009, 277 (2009)
97. Rührmair, U., Xu, X., Sölter, J., Mahmoud, A., Majzoobi, M., Koushanfar, F., Burleson, W.: Efficient power and timing side channels for physical unclonable functions. In: Cryptographic Hardware and Embedded Systems–CHES 2014, pp. 476–492. Springer, Berlin (2014)
98. Saha, I., Jeldi, R.R., Chakraborty, R.S.: Model building attacks on physically unclonable functions using genetic programming. In: 2013 IEEE International Symposium on Hardware-Oriented Security and Trust (HOST), pp. 41–44. IEEE, New York (2013)
99. Schapire, R.E.: The strength of weak learnability. Mach. Learn. 5(2), 197–227 (1990)
100. Schapire, R.E., Freund, Y.: Boosting: Foundations and Algorithms. MIT press (2012)
101. Schuster, D., Hesselbarth, R.: Evaluation of bistable ring PUFs using single layer neural networks. In: Trust and Trustworthy Computing, pp. 101–109. Springer, Berlin (2014)
102. Secure embedded systems (SES) lab at virginia tech: on-chip variability data for PUFs. http://rijndael.ece.vt.edu/puf/artifacts.html
103. Servedio, R.A., Tan, L.Y., Wright, J.: Adaptivity helps for testing juntas. In: Proceedings of the 30th Conference on Computational Complexity, pp. 264–279 (2015)
104. Servedio, R.A.: Efficient Algorithms in Computational Learning Theory. Harvard University, USA (2001)
105. Shalev-Shwartz, S., Ben-David, S.: Understanding Machine Learning: From Theory to Algorithms. Cambridge University Press, Cambridge (2014)
106. Siegenthaler, T.: Correlation-immunity of nonlinear combining functions for cryptographic applications (Corresp.). IEEE Trans. Inf. Theory 30(5), 776–780 (1984)
107. Suh, G.E., Devadas, S.: Physical unclonable functions for device authentication and secret key generation. In: Proceedings of the 44th annual Design Automation Conference, pp. 9–14 (2007)
108. Tajik, S., Dietz, E., Frohmann, S., Dittrich, H., Nedospasov, D., Helfmeier, C., Seifert, J.P., Boit, C., Hübers, H.W.: Photonic side-channel analysis of arbiter PUFs. J. Cryptol. 30(2), 550–571 (2017)
109. Tajik, S., Dietz, E., Frohmann, S., Seifert, J.P., Nedospasov, D., Helfmeier, C., Boit, C., Dittrich, H.: Physical characterization of arbiter PUFs. In: Cryptographic Hardware and Embedded Systems–CHES 2014, pp. 493–509. Springer, Berlin (2014)

110. Tajik, S., Lohrke, H., Ganji, F., Seifert, J.P., Boit, C.: Laser fault attack on physically unclonable functions. In: 2015 Workshop on Fault Diagnosis and Tolerance in Cryptography (FDTC). pp. 85–96 (2015)
111. Tajik, S., Nedospasov, D., Helfmeier, C., Seifert, J.P., Boit, C.: Emission analysis of hardware implementations. In: 17th Euromicro Conference on Digital System Design (DSD), pp. 528–534. IEEE, New York (2014)
112. Tarnovsky, C.: Deconstructing a 'Secure' Processor. Talk at Black Hat Federal 2010. http://blackhat.com/presentations/bh-dc-10/Tarnovsky_Chris/BlackHat-DC-2010-Tarnovsky-DASP-slides.pdf (2015). Accessed 25 Feb 2017
113. Tehranipoor, M.M., Guin, U., Bhunia, S.: Invasion of the hardware snatchers: cloned electronics pollute the market. IEEE Spectr. (2017)
114. Tobisch, J., Becker, G.T.: On the scaling of machine learning attacks on PUFs with application to noise bifurcation. https://www.emsec.rub.de/research/publications/ScalingPUFCameraReady/ (2015). Accessed 18 May 2015
115. Torrance, R., James, D.: The state-of-the-art in IC reverse engineering. In: Cryptographic Hardware and Embedded Systems-CHES 2009, pp. 363–381. Springer, Berlin (2009)
116. Van Tilborg, H.C., Jajodia, S.: Encyclopedia of Cryptography and Security. Springer Science & Business Media (2014)
117. Vapnik, V.: Estimation of Dependences Based on Empirical Data: Springer Series in Statistics (Springer Series in Statistics). Springer-Verlag, New York, Inc (1982)
118. Vapnik, V., Chervonenkis, A.Y.: On the uniform convergence of relative frequencies of events to their probabilities. Theory Probab. Appl. $16(2)$, 264 (1971)
119. Verbauwhede, I., Schaumont, P.: Design methods for security and trust. In: Design, Automation & Test in Europe Conference & Exhibition, DATE'07, pp. 1–6 (2007)
120. Xu, X., Rührmair, U., Holcomb, D.E., Burleson, W.P.: Security evaluation and enhancement of bistable ring PUFs. In: Radio Frequency Identification, pp. 3–16. Springer, Berlin (2015)
121. Yamamoto, D., Takenaka, M., Sakiyama, K., Torii, N.: Security evaluation of bistable ring PUFs on FPGAs using differential and linear analysis. In: 2014 Federated Conference on Computer Science and Information Systems (FedCSIS), pp. 911–918 (2014)
122. Yu, M.D.M., Devadas, S.: Recombination of Physical Unclonable Functions (2010)
123. Yu, M.D.M., Devadas, S.: Pervasive, dynamic authentication of physical items. Queue $14(6)$, 70 (2016)
124. Yu, M.D.M., Hiller, M., Delvaux, J., Sowell, R., Devadas, S., Verbauwhede, I.: A lockdown technique to prevent machine learning on PUFs for lightweight authentication. IEEE Trans. Multi-Scale Comput. Syst. $2(3)$, 146–159 (2016)
125. Yu, M.D.M., M'Raihi, D., Sowell, R., Devadas, S.: Lightweight and secure PUF key storage using limits of machine learning. In: Cryptographic Hardware and Embedded Systems–CHES 2011, pp. 358–373. Springer, Berlin (2011)
126. Yu, M.D.M., Sowell, R., Singh, A., M'Raihi, D., Devadas, S.: Performance metrics and empirical results of a PUF cryptographic key generation ASIC. In: 2012 IEEE International Symposium on Hardware-Oriented Security and Trust (HOST), pp. 108–115. IEEE, New York (2012)
127. Yu, M.D.M., Verbauwhede, I., Devadas, S., M'Raihi, D.: A noise bifurcation architecture for linear additive physical functions. In: 2014 IEEE International Symposium on Hardware-Oriented Security and Trust (HOST), pp. 124–129 (2014)

Printed in the United States
By Bookmasters